WATERTIGHT

Watertight

How I Survived The Submarine Service without Losing My Mind

A MEMOIR

KARL WILLIAM HECKMAN

Irvington House

IRVINGTON HOUSE PUBLISHING

Published by Irvington House Publishing, Portland, Oregon USA

Library of Congress Control Number: 2014906283

FIRST EDITION

ISBN: 978-0-9913-9850-8

Printed in the United States of America

The typeface of text and titles is New Caledonia in recognition of the book
Das Boot by Lothar-Günter Buchheim,
the First American Edition of which was set in the Caledonia typeface.

www.irvingtonhousepublishing.com

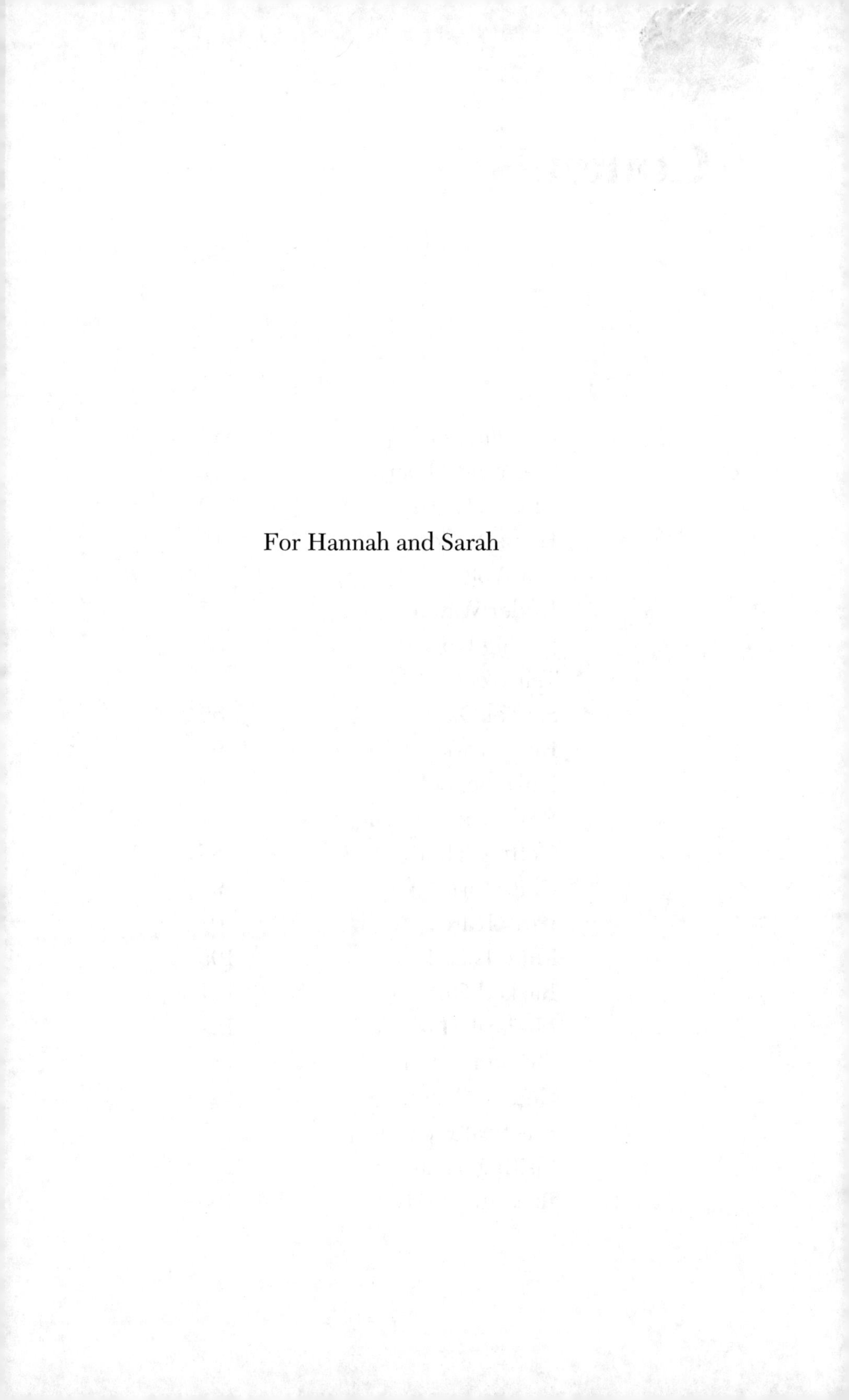

For Hannah and Sarah

Contents

Introduction

A SILVER INSIGNIA, a pair of dolphin fish guiding a vessel down into the sea, hung above the left breast pocket of the last Navy uniform I wore. Those dolphins designated me as Qualified in Submarines.

Years after my Navy time, submariners I served with stay in touch. Between us it only takes a word or a name to bring a chuckle. Those chuckles shift in timbre when recalling dangerous times.

As in: "The mission of '85…" "Oh, yeah, how could I forget…"

The equation determining life or death in a submarine is simple. If the number of surfacings does not equal the number of dives, your children become fatherless. The slimness of that margin for error kept us vigilant and demanding of our shipmates. A man lacking knowledge or ability is of no use to submariners and is expected to correct his deficiencies before he uses up oxygen meant for qualified men.

Within are my experiences; yet they illustrate themes common to all submariners past, present, and future.

All characters herein are real submarine sailors; there are no composites, no characters invented and inserted to add color. All color is actual. Conversations have been paraphrased for clarity and to clean it up for public consumption. Some names have been changed in this book to protect the innocent (mostly the author) from those who take exception to my memory of events. If you were a womanizing booze-hound, now reformed with loved ones who need not be aware of your past wicked ways, I have endeavored toward kindness and given you a pseudonym.

Let me also make an apology in advance to my shipmates: These are events as I recollect them. Scenes from my time in submarines remain in my head like snapshots in an old family album. Without Grandma's elegant cursive on the back of the photo there are always faces to which you can't match a name. Twenty-some years have passed since I left the Navy. Many technical details have slipped from my brain's ROM and RAM and into oblivion, please forgive my knowledge deficiencies. I did some fact checking with other old submariners; even so, any errors herein are mine alone.

Literary license is hereby claimed for situations where I wasn't present at all elements of an event. What I portray here is what was described to me by others who were there.

I'm still bound by an oath of silence regarding the nature of our missions and capabilities. Everything "classified" that I reveal in this work is already in the public domain.

The correct terms for Navy enlisted are "rate" and "rating;" rate meaning pay grade, rating referring to occupational specialty. Herein I occasionally interchange the term "rank," which while irritating purists, is used for clarity to a broader audience.

Also, I was on more than one submarine and several shore installations; it's possible that time has mixed up people or places in my mind. I remember dialogs and events well; sometimes I don't recall every man who was present at that moment. If I have written you out of a scene you were in, I apologize. If I have written you as doing or saying things you didn't do or say, please take that with good humor and rest easy knowing that I am confused without intent to misrepresent, and anyway, I probably wrote you to be more interesting than you were. (Zing! Those of you new to submarines will soon discover that a good insult is highly regarded; a great insult is revered.)

Thank you shipmates for your service in a dangerous undeclared and unaccredited war for information. Leaders, eras, ideologies, and causes come and go, but some things never change. As with all military personnel since mankind became uncivilized enough to pursue its interests through force of arms:

THEY SAID GO AND DO—WE WENT AND DID.

At 19 years of age when I enlisted in the Navy, most psychologists say that who I was then and am now had already been determined. Even so, I now realize that every day my actions, motives, thoughts, assessements, and judgements of situations, characters of others and decisions they make; my business decisions, my personal choices—even though they have shifted and evolved with more "time in life"—are all firmly rooted in my Navy experiences. If it's true that my formative years were past when I joined the Navy, perhaps my Navy time should be called my "tempering years." Whatever steel I was made of at 19 years old, during the next 9 years, 4 months, and 28 days would be heated red-hot, hammered, quenched, drawn, ground, polished and sharpened over and over until the me I am today emerged.

This book is an attempt to explain how the experience of being a U.S. Navy submariner gave me—and many others—a unique and lasting perspective of the world and the way we interact with it.

Fair winds and following seas, gents. Thanks for the memories.

Karl William Heckman
formerly ET1(SS)
USS Seawolf (SSN 575)
USS Parche (SSN 693)

www.WatertightTheBook.com

Sub-MARE-in-er
is a term best applied to a quality wristwatch or
to a poor sailor

"Submariner"
as used in this book is correctly pronounced
sub-mah-REEN-er

1

Welcome Aboard

You are starting a new life, with new surroundings and new friends. Grin and bear it like we all did.

— The Bluejackets' Manual, 1940 edition

THE ODOR JOLTED me. I leaned over the open hatch to see the ladder disappearing below and was enveloped by a column of stale air rising out of the submarine. I had heard rumors about the smell but also had learned to heavily discount wisdoms offered by fellow trainees. Military men are notorious for exaggeration; my five years in the Navy had schooled me that the man reluctant to self-aggrandize is the man worth listening to. Nevertheless, even bullshitters are right occasionally, and now, the first time I climbed onto a submarine, I learned that it was real—this olfactory insult known as "boat smell." Its suddenness unnerved me.

My brain said it was no big deal, only a twang in the air, like a breath of farm country. But my gut wasn't listening to my brain. My stomach knotted like when I was eleven and an older boy chased me off the high-dive at the public pool.

On looking down from the spring end of the diving board I changed my mind about jumping. The water was dark in the roped-off deep area under the high board and I'd already broken an arm and carried scars from a dozen stitches. My survival instincts sharpened by emergency room experience, I decided to skip the high-dive. But the tough kid wouldn't let me back on the ladder.

"There's only one way down for you, punk," he said. There was no way for me to know this event marked the beginning of a lifelong trend of choosing between undesirables.

Not a fighter, I executed what I imagined was a swan dive. I'd watched the Acapulco cliff divers on *Wide World of Sports* and knew that a diver was judged on how little splash he made entering the water. My hands held straight and flat pierced the surface like a knife. But the top of my head smacked the smooth water like a bowling ball on concrete. Stunned, I plunged deep into the blue water. The shock of the cold in the 12-foot deep end roused me enough to look to the surface where the air was. My eleven-year-old humiliation avoidance reflex powered my arms to jerk my swim trunks up from around my knees. The high-dive was exhilarating, bully or no.

Now, thirteen years later, the exciting path *is* the ladder. Once again I was beyond turning back. The future I had worked toward since joining the Navy was down this ladder.

My guide introduced himself as Bob Cranker. He would be my sea-dad, Navy-talk for a mentor, an experienced man assigned to a fresh report. He would show me around and help me get qualified on both the nuclear engineering aspects of the submarine as well as the boat itself so that I could earn the designation "Qualified in Submarines." I hoped he would guide me away from rookie mistakes.

I followed Cranker down the ladder to a blue-tiled deck seething with men in blue dungarees. We stood surrounded by gray paint and stainless steel.

"This is Stern Room Berthing," Cranker said. Bunks faced with blue curtains were stacked like archive library shelves.

"Most of the Nukes sleep here," he said. "The COB will assign you a rack, but you'll probably be back here with us."

"The cobb?"

"Chief of the Boat. Our Command Master Chief," he said. "Call him COB, one word, no letters."

Navy tradition referred to many titles by their initials or acronym or by something close enough in military-speak; CO, XO, LPO. The six school commands in my immediate past each posted a senior enlisted man, the Command Master Chief. He was referred to as "Master Chief." Now I'm told it's different for one guy. *Look out,* I thought. Hazing is high sport in the Navy and I was now in the fleet. I vowed to listen before I addressed anyone by an abbreviation.

Six men awaited their turn at the Stern Room ladder we had just descended.

"Up the ladder," they yelled at the thirty-inch circle of blue sky visible through the hatch. Topside, men waited to get in.

"Down the ladder, official business," thundered a figure started down. A blue uniform dropped down the ladder, skipping the last three steps by sliding his boondockers down the smooth steel handrails. He landed with a thump and a grin.

"Thanks, fellas!" he said.

"Bullshit on your bullshit official business, Meadows, you fuck," someone threw back at him.

Meadows laughed. Still grinning, he shook his head. "I can't believe I'm such an asshole." He was pushed aside as men darted up the ladder and disappeared.

In the submarine people brushed past each other by twisting sideways at the last instant. That three-foot circle I'd known out in the world as personal space compressed around me. My personal space was now defined by the fabric of my shirt.

Port side, against the outboard bulkhead, were two stainless steel doors. Cranker opened one to reveal a head. Instructions for operating the valves necessary to flush the toilet were posted on the stainless steel walls of the enclosure.

"When they blow sanitary tanks overboard they hang a sign on the knob," said Cranker. "But always crack the ball valve slow and watch for pressure before you slam it open. Some mean bastard might yank the sign on you. That'll change your life in a nanosecond."

It seemed to me that his eyes glinted a little too much as warned me. I filed that bit of information along with cautions about operating the head.

A closet on the other side of the passageway held a small sink, two washers, and two dryers. Cranker informed me that Saturday afternoon was laundry day for our division.

"Rain locker here," Cranker slapped the stainless steel door. "No Hollywood showers at sea. Get wet, turn it off, soap up, and then turn it back on and rinse. No bullshit, people will get mad if you're a water hog."

Cranker pointed to a small hatch cover in the deck. "After Aux down there."

"What's After Aux? There wasn't any After Aux at Prototype," I said.

Cranker laughed. "Prototype is the ass-end of a pretend submarine on land to train non-qual pukes. You're on a real submarine now, son."

He pointed at the hatch and explained. "After Auxiliary Machinery Space. Air conditioning plants and steering hydraulics and a bunch of other stuff. You'll learn all about it in your ship's quals."

I followed Cranker forward through a watertight door into the next compartment. The watertight door opening started a foot above the deck and stopped at chest height. Passing through required a syncopated step-and-duck routine. Like everything else on the submarine, the watertight door frame was hard steel. A careless move would severely bash shins and elbows, or dent a skull.

We passed the cordoned-off open door of the Maneuvering Room and moved up the narrow passageway. The long tubular shape reminded me of an airplane except here the passenger seats were replaced by machinery. My mentor moved forward, pointing left, right, up and down, rattling off facts quicker than I could follow his view.

"This is Engine Room Upper Level. Here we have Propulsion Turbines port and starboard, Reduction Gears port and starboard, and Ships Service Turbine Generators port and starboard.

Right above you are Main Steam Cutoff Valves MS-3 and MS-4, there's the crossover valve." Each valve hanging a foot over my head was the size of a beer keg.

"Around the corner there, port side, is the Secondary Sample Sink. All the way outboard starboard is Emergency Diesel Number One, Number Two is port side, and this is the diesel workbench. You'll spend a lot of time here when you qualify Reactor Technician."

A man wearing khakis looked up from reviewing a clipboard of machinery logs. The corners of his eyes crinkled.

"What do you have here, Petty Officer Cranker? An FNG?" he said.

Cranker introduced him as Chief Lethin, Electrical Division Leading Chief Petty Officer, on watch as Engineering Watch Supervisor. The chief shook my hand.

"Welcome aboard," he said. "What was your last duty station?"

I told him I had just come from the S1C prototype in Connecticut.

"The Bay Area isn't the warmest spot on the West Coast, but we don't get those goddamned New England winters," the Chief said. He surveyed my uniform. "Petty Officer Second Class with no dolphins," he said. "Were you a staff pickup?"

From each prototype graduating class a few students were selected to stay on as instructors—staff pickups.

"Yes, I was," I said. "But don't worry Chief; I'm not arrogant about it."

"We'll see, won't we?" he said. Smiling he added, "Get hot, get qualified, and we'll get along fine."

Cranker led me forward through another watertight door. Before I could straighten to height, Cranker continued his rapid -fire tour guide routine. More pumps, piping, and valves, heat exchangers, air vents, and rows of six-foot tall electrical cabinets. Cranker stopped at a squat cabinet.

"This is the Reactor Compartment Air Particulate Detector. It's a piece of shit and you'll learn to hate it." He laughed, "I promise, it already hates you."

We stepped up to a circular walkway surrounding a tall stainless steel cylinder in the center of the compartment. Cranker slapped the stainless steel and announced, "This is the top hat. We are standing right above the reactor vessel. The control rod drive mechanisms extend up through the deck into this. This is part of the secondary shield. Over there are the Nuclear Instruments, the Primary Plant Instruments there, Protection and Alarm system here."

Several men stood around a small cabinet serving as tool storage, workbench, desk, and meeting place.

"And this," Cranker gestured a mocking flourish, "is Reactor Controls Division." The men introduced themselves. I already knew Cranker, now I met Jim Boivin, Mike Plant, Mike Ouellette, Rick Traver, Mike Bernhardt, and Ken Kohl. Mark Scansen was on watch in Maneuvering.

They had previously worked out that there were too many Mikes in the division. Mike Plant retained his given name, the other two were known as Omelet and Bernie.

They asked where I had most recently come from, if I was married, kids, the whole thing. Married, yes. Kids, yes. Two small daughters; one walking, one working up to crawling.

One of my new coworkers asked if I had attended LMET. Leadership Management Education Training was part of the "new Navy," a course on effective leadership beyond traditional command and control. It focused on understanding people, motivations, Maslow's Hierarchy of Needs, and trust building. My yes brought guffaws and snorts.

"Don't try any of that touchy-feely bullshit here!" one said. All laughed.

"Yeah, there is only one motivational tool here," another quipped. "Fear," he said at the same time another man said "Death." All laughed.

"That's right, fear of death," said Kenny Kohl, the Leading First Class Petty Officer of the division. "It's the sharpest tool in my toolbox. But I've still got the hammer." All laughed again.

A heavy footfall on the deck plate under the reactor compartment forward watertight door announced the entry of the Reactor Controls Division Leading Chief Petty Officer, Senior Chief Electronics Technician Johnson.

He'd never laid eyes on me before now, but he glared at me and said, "Aren't you qualified yet?" The rest of the division laughed.

Petty Officer Ouellette spoke up. "He's close enough. I stand relieved!" Even the senior chief chuckled. These guys liked to laugh. This may not be so bad.

The senior chief quizzed Cranker. They spoke in verbal shorthand.

"His qual cards?"

"Almost ready. First thing in the a.m. Senior Chief," Cranker replied.

"Ship's quals?"

"Same, same, first thing."

"Good. We are going into dry dock next week. I want Petty Officer Heckman to set a record qualifying SRO," the senior chief said. He was looking directly at me, the humor gone from his eyes.

"You will qualify as Shutdown Reactor Operator in record time, won't you Petty Officer Heckman?" There was only one response to this rhetorical question.

"Yes, Senior Chief."

2

Charlie Oscar

Bullshit may get you up the ladder, but it won't keep you in the tree.

— *The Submarine Sailors' Imaginary Book of All Knowledge*

CRANKER SHOWED ME other parts of the submarine; the Crew's Mess, Control, the Projects area, and forward to the Torpedo Room. He introduced me to people whose names I forgot in the haze of names and faces I met that first day. I met men from other engineering department divisions: Machinist's Mates Rob Kersch and John Parrott, Electrician's Mates Keith Lyly and Clarence Davis, and an Engineering Laboratory Technician Jim Stanton. These guys seemed friendly yet slightly wary, not quite committing to friendship at this point. They were waiting to see what brand of man I would turn out to be.

Back in Reactor Compartment Upper Level talking with my new division, the deck plate on the Reactor Compartment side of the

Control Room watertight door rattled again. A man ducked through the hatch combing and straightened. A trim khaki uniform pressed sharp and straight stood tall. Silver eagles adorned his collar points. This could only be one person—the captain of this submarine.

A red mustache swept past the corners of his mouth, seeming out of specification with Navy Regulations on mustache grooming.

Yesterday I had been reporting to an office. In a busy-work program with other sailors awaiting the submarine's return from sea, my military appearance had slipped into California mode. My uniform was crisp, but I had let my mustache grow out of regulation. It was carefully trimmed but longer in all directions than allowed. I wished I had been less free-spirited in front of the mirror this morning.

The epitome of command presence, the Captain exuded authority. His face quiet and his demeanor calm, he looked like a man who knew his business. There was steel in his unblinking gaze.

Reactor Controls Division greeted him warmly. While professional and respectful, they exhibited no hint of the false uber-courtesy that afflicts lifers and kiss-asses in the presence of their superiors. At previous commands I'd been at, people started twitching whenever "Charlie Oscar" was in the area. Officers often "formal-ed up" to a crowd of enlisted, referring to everyone as Petty Officer So-and-So, in a concerted effort to get credit for showing respect. That usually came off as obvious condescension. Less often but equally as pathetic was an officer trying to suck up to the enlisted folk with an unskilled 'I am of the people' routine. Most Navy officers didn't have much smooth. There's a reason these guys weren't in politics or on television.

No matter what words came out of their mouths, officers' arm's-length demeanor made it clear: You and I are different, and I am better. You wait for your turn to eat from plastic plates and use a paper napkin to swipe gravy from your chin; I dine from china, a silver ring engraved with my name holds a linen napkin at my place at the wardroom table.

The Captain was different. His competence and authority were obvious; he had no need for displays of superiority. His conversation with the crew came unforced. These *Seawolf* men were comfortable around their commanding officer, and he around them.

The Senior Chief introduced me as the new guy. The Captain inquired if my family was settled in, did we need anything? Impressed with his interest, I relaxed until his eyes fell on my mustache. Hands on his hips, he studied my upper lip for a moment.

My stomach churned. I wanted to make a good impression, but now…

"Nice 'stache," he said. Then, "Carry on, gentlemen," and he strode toward the Engine Room.

Relieved and surprised, one thought rose above the cacophony of new information roaring in my head: I'm going to like it here.

3

Horse & Cow

A good bar is a sacred place where one is free to contemplate past poor decisions and make new poor decisions, all without fear of judgment.

— *The Submarine Sailors' Imaginary Book of All Knowledge*

THE MEN OF Reactor Controls division were finishing up their days' work of maintenance, training, and its associated paperwork. Being my first day, I was unqualified to do anything helpful. Eight men stood, sat, knelt, and leaned in a workspace the size of my apartment kitchen. I shuffled out of one man's way only to be in another's. The Senior Chief had gone away to do senior chief things. The Senior Chief was retiring in a few weeks and I wondered if he was off on Navy business or was on the job of lining up his retirement ducks. The Leading First Class, Kenny Kohl, was the next senior man in the division. Like all men one rung from the top, he handled the lion's share of the routine day to day management of the men.

"Listen Heckman," Kohl said, "nothing personal, but you're of no use to me here. Go home; be at morning quarters tomorrow at 0700. That'll be topside." He pointed an unwavering finger at me for emphasis. "Do not be late."

"Aye, aye," I responded. Kohl added a post script.

"The only acceptable reason to be late for morning quarters is a death in your family," he said. "But there's a catch to that."

"Yeah?"

"The death has to be your own.""Transmission received," I said. "I'll be there."

Grateful for the reprieve I gathered my things.

Kohl turned his head to look at me out of one eye. "Would you by chance be a drinking man?" he queried in a put-on Irish accent. I was a babe on this boat, but I'd been in the Navy five years. I knew a kindred spirit when I heard one, even through a fake accent.

I responded to the question with as innocent a face as I could. "I've been trying to learn how to drink beer," I said. "Maybe you guys could give me some pointers."

"Ah…" said Kohl. He turned to the men nearby. "Gents, this one is a bad liar." He tapped his temple and with a wry grin said, "Good to know."

Those men not staying for overnight duty were going to their favorite bar after work and I was invited. I could get acquainted with my new coworkers and get a jump on things I needed to know. The informal grapevine was often quicker and more accurate than formal channels of information.

They gave me directions and sent me on my way.

At 3:30 in the afternoon it appeared to be a standard dive bar; it probably looked better at night. A concrete block held the front door open.

Inside, one large room faced me. To the right, a long bar ran along the wall. Three pool tables crowded the rear of the room. To the left lay a dance floor. On closer inspection, the dance floor flaunted its previous life as a bowling alley. Arrows and dots were visible under worn varnish. A mirrored disco ball hung over the dance floor. Wooden beams supported the ceiling. On the sides of those beams, and on the walls, and on back bar shelves, hung an array of submarine memorabilia surpassing the submarine museum I had visited in Groton, Connecticut.

The museum in Groton displayed portentous artifacts like the wardroom silverware from *Nautilus*, the first nuclear submarine. *Nautilus* officers' eating utensils were fashioned of Zircalloy, a special alloy of the rare element Zirconium invented under the direction of Hyman Rickover as the best material to clad nuclear fuel plates.

Groton's submarine museum exhibited ships bells and commemorative plaques hung alongside posed photos of crews and submarines, many rendered in black and white. Photos of submarine launchings complete with invitations sent to VIPs, and handshakes with Presidents where solemnly framed and hung in neat rows. Displays were staid, austere, and officially sanctioned. They had been donated in response to courteous letters on embossed stationary, which were also displayed.

But in the Horse & Cow bar in Vallejo, California the items were different. Here was contraband; stolen mementos and pilfered parts. The Horse & Cow was a crime scene. Submarine sailors had liberated odds and ends from their boats and nailed those bits to the Horse & Cow.

Banners with submarine names and mottos were strung on walls. Hardhats stolen from shipyard supervisors sat on the back bar. A brass ship's bell was screwed to an overhead beam, a knotted cord dangled from its striker within reach of the bartender. Torpedo tube label plates hung on wires, the entire stainless steel door of a head leaned in the corner. A reactor siren, reactor warning lights, and a reactor scram switch projected out of a wooden plaque.

A sign suspended from a braided wire loop was engraved with block letters:

DO NOT USE HEAD
WHILE BLOWING SANITARY TANKS

A stainless steel submarine toilet seat hung on the wall like a lucky horseshoe. A plaque positioned in the middle of the toilet seat was lettered “Executive Officer’s Stateroom.”

Names of people and submarines were carved into the wooden elbow rail on the bar top. Here and there could be seen a carved heart with a woman’s name inside. I saw a heart with two names, a man’s and a woman’s, with a large X carved through the whole thing. The X looked fresher than the original lettering. The second cut was deepest.

On the back bar sat a clear plastic keg with a tap. A light positioned underneath shone into the jug, causing the boozy concoction inside to glow green. The keg was labeled “Nuke Waste.”

I sat on a barstool and pondered why only a few tables and chairs were available in a room so cavernous. Perhaps patrons didn’t sit here, or maybe the chairs had been smashed over heads in epic bar fights.

Music played from somewhere, too loud for the empty room. The bartender was an older gentleman who looked as though he had been around the block and seen it all. Sporting a thick shock of graying waves, he could be forty-five years old or sixty-five, I couldn’t tell.

He offered a handshake and said “I’m Jimmy. I own the joint, but don’t hold that against me. I was brought up bad.”

Jimmy asked what I wanted to drink; I asked what kind of beer he had. He named brands, counting them on his fingers. He listed one that sounded perfect for this warm spring day. Jimmy, I’ll have a Michelob, please.

He retreated to a cooler and returned with a bottle. Snapping off the top, he placed before me a bottle of Olympia beer.

It wasn’t what I had asked for, but I drank it anyway. It was a little loud in here and Jimmy is not over-young. He probably heard wrong. I had no worries about the brand; any beer would do, so long as it was cold.

Still curious I had to ask; “Hey Jimmy, why so few chairs?”

“Do you know what those fuckers cost?” he responded. “God-damn straight-up robbery.”

I was startled by the roar of a motorcycle engine. The rider revved his engine and goosed his bike over the tall curb onto the sidewalk. He rode through the front door, dropped the kickstand in the middle of the barroom, and with a final wrap of the throttle shut off the engine. Dressed in worn black leathers, a Harley logo across the back of his vest, he looked like a recruiting poster for a down-at-heel motorcycle club.

Climbing off the bike, he moved to a barstool. With the high-brow courtesy of an English aristocrat, he said, "May I acquire a beverage, my good man?"

Jimmy moved to the cooler. With an almost clean bar towel draped over his forearm he displayed the beer bottle label for approval. "My finest vintage, Sir Fuckhead."

"Just give it here, you old fart," Harley Man laughed. Jimmy placed the bottle on the bar. As the man reached for it, Jimmy slid the bottle sideways out of reach. Both men laughed.

Harley Man looked past Jimmy at the back bar with sudden concern. "What the hell…?"

Jimmy swiveled around to look. Harley Man snatched the beer bottle off the bar.

"Thanks Jimmy."

"You're very fucking welcome," said Jimmy disgusted at being tricked so easily.

My beer bottle had been motionless halfway to my mouth for the last minute as this bizarre event played out. Jesus, I thought. This is a psycho ward. My new coworkers — my new buddies, my pals — must have set me up. They're at another bar pounding down cold ones, laughing at me for being here. This place is not normal in any sense of the word.

Jimmy came down the bar. "Ready for another?"

"Sure," I said. "This time I'll have a Michelob."

"Michelob, coming right up," Jimmy said. A moment later a fresh Olympia was on the bar in front of me. I knew he heard me correctly, he repeated my order for a Michelob. What the…?

Looking down the mahogany at my only bar mate, I saw an Oly in front of Harley Man. Finishing his beer, he slapped money on the bar and clambered off the stool. He tossed the empty beer bottle into a metal garbage can twenty-five feet away. The bottle slammed home with a bang and a rattle.

Harley Man threw a leg over the saddle of his bike.

"Thanks, Jimmy," he yelled as he fired up the engine.

"You're welcome, asshole," Jimmy yelled back. "Now get that stinking thing out of here." Harley Man smiled, waved, and rode out the blocked open back door, revving his engine all the way. The staccato roar faded into the distance.

In the refreshing silence, Jimmy shook his head. "Some people's kids…" he lamented.

When I asked about the beer selection, Jimmy said, "I only carry one brand. It makes things a lot easier." He bought whatever brand the distributor was selling cheapest. The brand changed occasionally, but only one brand of beer was sold at a time.

That did indeed make things a lot easier.

4

Sea Wolf

Nature is not as forgiving as Christ.

— *Hyman G. Rickover*

Seawolf MANAGED TO be unobtrusive even when tied to the pier in plain view. Three hundred-eighty seven-feet long and twenty-eight feet at the beam she seemed too large to hide in the open air. Nonetheless, until close upon her, the submarine blended into the industrial background along the shipyard waterfront.

Proper nomenclature classified her as a "ship," but among the crew she was simply "the boat." Home and haven to a hundred twenty-five men, she was simultaneously respected and despised. She allowed them to exist in an environment unnatural to their human bodies. She might also kill them in spite of their endeavors otherwise.

Past her twenty-eighth birthday, *Seawolf* was a senior citizen in a career field where thirty years marked a long full life. Her aging suite of machinery often malfunctioned, sometimes catastrophically, as old equipment run constantly will. The sailors who operated and cared for her said you needn't run emergency drills on *Seawolf*; she ran drills on you.

The most immediate feature was her sail. It appeared as a six-foot thick, twenty-five-foot high steel wall rising out of her spine a few feet above the water line. Mostly rectangular, longer than tall, I imagined the sail as a shark's dorsal fin. The sail housed periscopes, antennas, and other extendable masts useful to the submarine.

When surfaced, ninety percent of the submarine remained below water. Moored fast with topside hatches open to the sky, the men working, sleeping, taking their meals, showering, and washing their clothes inside her, were below the level of the river outside. Even in port they were subsurface.

It was as if the submarine was more at ease unseen. Where she had been, what she had done, what she was getting ready to do, were questions the answers to which were rigorously guarded. Access to classified information carried an additional requirement; the need to know. Most of the men on the ship did not have the need to know.

In the past few years the Soviets had been exceptionally lucky in deducing U.S. submarine capabilities and activities. Causes of that problem were not yet clear. Information leaks could come from a myriad of sources.

An apartment in Vallejo with a view of the river and the shipyard might also hold a binocular or telescope. The friendly guy at the Horse & Cow who bought beers for sailors, the nice young lady admiring Navy men, the older man at church who asks how long you'll be gone, all could be working for the other side. Or just one of them might be. Or none. How could you know?

If the submarine possessed eyes, she surely would have been watching to see who was paying too much attention.

In fact, she did have eyes. They were in the head of the sailor assigned as topside watch standing guard day and night. He was paying attention to who was paying attention.

One age-old method to ensure secrets were kept was to limit the number of people who knew. This protocol included the men actually accomplishing the task. They may know what they are doing, but not where. Other men knew where, but not what. Fewer knew why. We Nukes knew none of the above. We knew only the boat.

The lines of her hull were long and tapered yet graceless. Only the second nuclear submarine, *Seawolf* was built during an in-between era of submarine design. The engineering expertise amassed on submarines from their inception had been eclipsed by the development of fission as a power source. Fission for power grew as a stepchild of the Manhattan Project only a few years previous. Atomic power changed everything for submarines.

The submarine of the World War II diesel boat era was essentially a surface vessel with the capability of submerging for short periods. Nuclear power flipped that old standard on its head. With the advent of high pressure, high temperature steam propulsion made possible by the heat generated by a nuclear reactor, submarines could spend all of their time at sea submerged, surfacing only when close to port. The submarine stealth factor rose like a rocket from Cape Canaveral.

Fission did not require oxygen as did the diesel engines used previously for surface transit and charging of the ship's storage battery. For the diesel boats, the battery alone provided power to the electric propulsion motor during undersea operations. Its charge was limited; recharging required the diesel engines to turn a generator. Running the diesels required air.

Although nuclear submarines still carry a large ship's battery, a submarine power plant does not rely on the battery during normal operations. Also, the submarine's nuclear power plant charges the battery while submerged. No longer needing a daily dose of air to charge batteries allows the submarine to part company with the surface for extended periods. In 1960, the nuclear submarine USS Triton circumnavigated the globe in eighty-four days. Triton retraced the route of Magellan with one spectacular difference. Triton completed the distance submerged.

In place of diesel fuel, which required regular replenishment, the solid fuel loaded into the reactor core of the day had a useful life of 15 years or more. Not needing constant refueling gave submarines a range limited only by the quantity of groceries they could carry for their crews.

The fission chain reaction furnished far greater horsepower to the propellers than could the electric motors driven by the battery. A WWII diesel-electric submarine crept along while submerged. It was now possible for submarines to go faster submerged. This compelled changes in hull design to accommodate greater speeds and depths. Hulls became rounder in cross-section and sleeker in profile.

Seawolf's hull design was improved but carried over elements reminiscent of diesel boats. These elements included a superstructure—a covered framework supporting a flat wooden deck along the top of the rounded hull. Made of teak, this weather deck was coated with coarse sand to help a shoe grip a wet deck. Newer submarines would be designed without the superstructure and without wooden weather decks. *Seawolf* carried both.

Her weather deck sloped up to a blunted bow the crew called the bull nose which housed sonar gear. Beneath the bull nose the sides of the hull came together into a sea parting wedge.

Like a teenager tortured by her awkward years, *Seawolf's* exterior displayed the ungainly irregularities of changing needs and evolving design. Topside on her bow was a projection with the appearance of an upturned trash can, mounted and painted. Farther back another projection stood two feet tall like a telephone pole broken off at the knee. Farther aft the hump of a late-addition watertight hatch broke the continuity of the deck.

Flat black paint masked the ships age at a distance, but up close accentuated it, like a "past her prime" woman's face-lift that fails to remedy her sagging neck.

Behind her aftermost hatch, mounts for a Deep Submergence Rescue Vehicle angled upward. Flat on their tops, they looked like the bare ribs of an abandoned construction project. The DSRV was no longer attached yet the mounts remained.

This DSRV was also a Special Projects security misdirection; when I asked how the DSRV had operated, peoples' eyes glinted. It hadn't been a Deep Submergence Rescue Vehicle at all. It would take a while for me to figure out the truth of that system.

The first rule of a cover story is that it must be plausible. Give the media a solid bite of the apple and they won't notice that they have, in fact, a mouthful of pear. The public will believe that what the media tell them is true.

Those few citizens inclined toward skepticism will use the same media as sources and quote them to verify the story to their non-believer friends. It's a perfect circle and beautiful in its simplicity. It's so easy it ought to be a crime. You don't have to lie: just give up a small truth and no one will notice the bigger obfuscation. The public will suss out the undisclosed facts surrounding the small truth and celebrate their victory over government lies. Give them 'it's a Deep Submergence Rescue Vehicle' and they go away sated.

On initially arriving to the boat, one of my first stops had been a security briefing in a high-security conference room known as the Vault. The Petty Officer briefing me explained that everything having to do with the Projects Department was none of my concern.

I was only beginning to learn about secrets. *Seawolf* was like a department store filled with shelf after shelf of plain black boxes labeled Authorized Personnel Only. This store of secrets had an elevator with control panel buttons to floors requiring an encoded access card which I didn't have. I was not Authorized Personnel.

Forgive me — I'm jumping ahead…

Over the submarine's nearly three decades of service, experimental equipment of various kinds had been welded on and later cut off. This left scars and welded patches. And then, of course, was *Seawolf's* biggest secret, what she might describe to her girlfriends as her "little procedure."

Formed in the mid 1960s—the Cold War demanding surreptitious diligence—the Office of Special Projects needed a special submarine for a special purpose. Being the only one of her class, and not to be duplicated, *Seawolf* was a perfect choice to develop into a specialty. The same could be said for her predecessor *Halibut*. Uniqueness didn't lend itself to fiscal efficiency in Henry Ford's production line, nor in Hyman Rickover's nuclear navy. An orphan boat type was a problem for spare parts and upgrades; one might as well go whole-hog and make the old girl a truly one of a kind platform. Marty and Doc Brown didn't go *Back to the Future* in a Dodge Dart, now did they?

Reference material on warships such as Jane's Fighting Ships list the 'Wolf's length differently depending on the date of the research. Her keel was laid one length, but now was something different. I was aware that *Seawolf* had spent a couple of years in dry dock being "stretched," but didn't yet know the what-for and why.

I had asked about the large dark holes visible just above her waterline at either end of the boat. They seemed odd to me. The topside watch I asked fixed a steady gaze on me.

"I don't know what you're talking about," he said. I pointed to round grated openings in the boat's superstructure.

"Those," I said. He did not look where I pointed; his eyes remained locked hard on mine.

"I don't know what you're talking about," he repeated.

Oh, okay—I get it. I shut up and went on my way.

The submarine was held to the concrete pier by six lines. Two wooden rafts of rough timbers, each ten feet wide, floated between hull and pier. These "camels" kept the boat a safe distance from the pier to protect the submarine's diving planes and propellers. The diving planes were short wings projecting out of the boat on both sides. Bow planes were located forward, stern planes aft. They rotated hydraulically to control the boat up or down when submerged.

A portable walkway—a brow—bridged the water gap between pier and boat. One brow was set on the forward half of the boat and one aft.

The world was still dark when I came in to work this morning. The forward brow was lined with a string of bright lights zip-tied to the handrail. Each bulb had a shield to direct the light downward to the walkway, but two were twisted up. Raising a hand to ward off the blinding light and turning my head away, I saw things had changed since I left just a few hours ago.

During the night shipyard support elements had been stripped away in anticipation of the ship's departure. The aft brow had been removed and the shore power electrical cables along with it. This meant the reactor was critical and producing heat. The steam plant was steaming. The submarine's electricity was being supplied from within by at least one of the two Ships Service Turbine Generators. The engine room was alive and the boat self-sustaining.

Several men stood topside, waiting at the stern room escape trunk with the resignation of men accustomed to waiting. The hatch and the ladder beneath it accommodated one person at a time, and the hatch was doing great business this morning.

An hour ago I hugged and kissed goodbye my wife Robin, little Hannah, and sleeping baby Sarah. I was in the Navy when Robin and I met. Warned that a day would come when I would sail away for an unknown term, she chose to share the path anyway.

Today was the day.

My excitement with the adventure was countered by a dull weight in my chest. I hated to leave my girls. It felt like abandonment, and I—husband and father—the abandoner. Yet I was bound by obligations made before I knew Robin and before the girls came to be. I had wanted to be in the Navy. Navies are seagoing enterprises, and seagoing means leaving.

When my turn came, I lowered my sea bag through the watertight hatch and followed it down into the boat.

5

Under Way

Under way: Said of a vessel when she is not made fast to the ground in any way.

— *Naval Terms Dictionary*

MY BUNK IN the stern room berthing area was slightly roomier than a coffin. A four-inch-thick mattress lay on a hinged pan which tilted up to reveal a storage space as long and wide as the mattress and four inches deep. I placed my personal belongings inside and laid a folded dress blue uniform between mattress and pan. The weight of the mattress, and me, would keep the creases sharp. I could not imagine needing a dress uniform on this trip.

Since this was my first time out I watched others for cues. Trained to operate a nuclear power plant, my experience until now had been sea-simulated. Sure, I had hundreds of hours operating a reactor on land, but this was the Navy. Whatever I had done on land did not impress submariners. They didn't give a fat rat's ass about my grade point average or my class standings. They wouldn't have cared if I had designed the reactor or first split the atom with a wooden hammer. The professional successes that mattered to them happened only at sea.

I had been in the Navy five years, but to the submarine qualified men of *Seawolf*, this was my first day.

I was ready but anxious. I knew my part in this production but had never experienced the ship in motion. How different would that be? Would I get seasick? This morning Doc had walked through the boat handing out Dramamine tablets. I declined. I didn't think seasickness would be an issue of mine.

Experienced men went about their business routinely, except with a little more hustle in their step this morning. Their eyes were bright; they cracked jokes and traded insults. A few were more hung-over than usual. Their last night ashore was not wasted on sleep.

At 0800 everything was ready.

Topside, the order rang out: Take in all lines.

A rookie of the Engineering Department, I was not yet qualified in at-sea watch stations. Of no use to my peers, I was unemployed while the other guys did their work to get the ship underway. Roaming the engineering spaces I watched and learned, and tried to stay out of the way.

People on the dock saw what I could only imagine. Let go from the pier, six mooring lines splashed into the Napa River. Pulled hand over hand onto the weather deck of the submarine, the lines were coiled and stowed in lockers built into the superstructure.

Below decks, the ships announcing system reverberated with the matter-of-fact voice of the Officer of the Deck: The ship is underway.

The boat jerked sideways and my heart rate spiked. My experience of the submarine up to now had been as an immovable solid object, an oddly shaped building where I worked. Now it moved.

A harbor tug eased *Seawolf's* black hull sideways into the channel of Mare Island Strait, released her, and backed away with a roar of diesel engines. Clouds passing above the open Engine Room hatch offered a gauge of our motion.

In Maneuvering the Engine Order Telegraph rang sharply and its black pointer advanced. All ahead, one-third.

The Throttleman spun his wheel, the low hum of the idling Main Engines climbed smoothly into a powerful whine. The surface of the river behind the submarine roiled as the propellers churned the water below. As the propellers gained purchase, *Seawolf* shuddered as if shaking off the last remnants of *terra firma*. She moved forward, slowly at first, gaining speed as inertia gave way to momentum.

"Here we go," smiled Rob Kersch. Rob was a good-natured Machinist's Mate from Montana. His quiet cheerfulness was reassuring. It would take a heap of trouble to sink Rob's buoyant spirit.

The clouds flew past the circle of sky I viewed from underneath the Engine Room hatch. I'm sure there was a grin on my face. I tried to appear nonplussed, even bored, to promote my image as a steady hand. I hoped I was looking calm and relaxed. Inside I was jumping up and down like a five-year-old on Christmas morning.

Everyone was busy. Everyone, except me. I asked Kenny Kohl, one of my many bosses, my division's Leading First Class Petty Officer—my Leading First—how I could help.

He stood in Reactor Compartment Upper Level watching the meters of the Primary Plant Instruments monitoring parameters in the reactor plant. Without shifting his gaze he said, "Tie a string, really tight, across the boat so you can see it go slack when we get deep." He glanced at me, grinning. There was a train-load of crap head games that old submarine guys played on new submarine guys. This squeezing of the boat at depth was a standard. Later I would find out it was true, but for now I communicated my disbelief submarine-style.

I faked a cough to thinly cover a mumbled, "Fuck you very much." My predictable insolence didn't faze Kohl, but something in my tone caught his attention. He looked closely at my face.

With an eye wise for his twenty-four years he tapped my chest and said, "Just keep breathing in and out." I took a deep breath and became a new man.

South, down the Napa River, west into San Pablo Bay, then south again into San Francisco Bay, USS Seawolf, SSN-575 steered for

California's Golden Gate and the blue water of the Pacific. Each turn reverse banked, the ship heeled over—leaning out—away from the turn. Once submerged, the motion would be opposite, as the boat leaned into the turn like an airplane. Submerged, the submarine "flew" through the water; surfaced she was awkward.

The old salts put on leisured airs. In truth, their attention was intensely focused on the machinery. When an equipment warning illuminated, or an alarm sounded, they reacted instantly. If the sound of a motor, pump, turbine, or any other piece of rotating machinery changed in pitch or volume, even slightly, they were there; monitoring, checking, attending-to.

Sailors hovered over machinery recently overhauled in the shipyard. They monitored temperatures and pressures, voltages, speed, everything that indicated the correct operation of the equipment. They watched for indications they knew to be trouble. *Seawolf* was their ship and everything must be right. It was their ass on the line. The sailors would have to deal with any malfunctions—at sea—while the overpaid shipyard union types slept safe at home on pillow-top mattresses next to their wives or girlfriends.

Neither frenzy nor wasted motion was present, only the practiced skill of professionals taking their ship to sea.

Word came down from the bridge that San Francisco Bay was breezy today with a light chop. The boat slowed as the bell was reduced. We heard that sailboats were crisscrossing our path like cocky teenagers as we passed Angel Island. Perhaps they didn't know that we would not stop or veer off to avoid them. Maritime rules of the road grant right of way to the larger vessel. Four thousand tons of nuclear submarine cannot stop on a dime. But never mind the physics—a warship does not suffer fools. Soon we throttled back up to speed.

An order passed down from the bridge. Come right to two-five-zero degrees. In the Control Room the helmsman steered to the new course. Watch standers on the bridge saw the boat's dark hull carve a white arc through the blue water until the bow pointed toward the flat horizon between two points of land; Fort Point on the San Francisco side and Lime Point, the southern tip of the Marin Peninsula.

The ship's announcing system speaker popped signaling its microphone being keyed. Engine Room conversations paused as men cocked an ear. Passing under the Golden Gate Bridge. Shortly we would cross the outermost limit of San Francisco Bay, a line between Point Bonita and Point Lobos—land's end. Beyond, the Pacific stretched over the horizon and covered one-third of Earth.

Crossing the line it was as if we'd hit a wall. The friendly waves of the protected bay were left behind. The full force of the ocean swept up the continental slope and leaned in on us unabashed by any land mass. The Bay had been a stroll in the park; there was power behind the heaving of the open sea. Currents circulating since the world began combined with forces of the moon to produce a landward rush of water twice each day. Enough water ripped through the three-quarter-mile wide Golden Gate to raise San Francisco Bay five feet. This morning the miracle was occurring in our face.

All ahead, Full. Propulsion turbine throttles opened, pouring high pressure steam onto turbine blades. The Main Engines rumbled as their tachometer needles spun clockwise. The noise level in the engine room rose. So did the temperature. Topside hatches had been shut before we left smooth water. Heat radiating from steam piping and components was ours to deal with. Piping and machinery was insulated, but that only slowed the transfer of heat. Engine room watch standers rolled up their shirt sleeves and shifted the seawater-cooled air conditioning plants to high speed.

Seawolf charged up the swells and crashed down into the troughs. I rode the rising deck then grabbed for a handhold as the deck dropped beneath me like a maniacal carnival ride. The deck, and everything in the boat, followed the hull; tilting up then nosing down into the next trough. Straddling a surfboard, this riding the waves would seem natural but in the 387-foot long submarine it was as if the world was out of control. The motion repeated, jerking and jolting like a Ferris wheel with loose gears.

Just when I had the rhythm a sneaker swell doused the bow between beats and the timing changed. This motion became diabolic when the boat angled into the waves. Then, as *Seawolf* rose and fell fore and aft, she rolled side to side. Everything twisted, including my stomach.

Inside the boat no horizon was visible for reference. Everything moved as one; deck, machinery, bulkheads, overhead. My eyes said all was normal. My inner ear disagreed. I tried not to think about it.

Each time *Seawolf* bottomed between swells, she shivered and sounded a loud boom.

My god, I thought. We are four thousand tons of mankind's most marvelous technology, tossed around like a child's toy in a bathtub. It was a sobering thought. Our journey was only beginning.

A bookshelf crammed with heavy technical manuals sat above me on the opposite side of the passageway. The bookshelf was rigged for sea, a metal strap kept the books on the shelf, but it didn't look sturdy enough. Each metal-hinged manual weighed enough to permanently dent my skull. I moved.

I settled into a new perch near the diesel workbench when Kenny Kohl sauntered by. Looking at me, he shook his head. "What did we just talk about?" he shouted over the racket.

"You mean the 'keep breathing' part?"

"Right." Another deep breath relaxed my grip on a pipe.

The Main Engines howled as the propellers lifted halfway out of the water then groaned as the stern dropped back into the sea and the screws bit deep.

I checked the reactions of more seasoned sailors hoping to gain confidence that this was normal and that all was well. The old salts were going about their work with disdain for the discomfitures. They timed the rolling deck and moved with it, finding something to grasp onto a second before the boat bottomed out with another teeth-rattling slam. Fresh curses flew with every crash of improperly stowed gear banging on the hard deck and much softer crew.

Bill Greiner, the Electric Plant Operator on watch, surrendered his breakfast into a black plastic trash bag he carried tucked under his belt in anticipation. Word traveled quickly, bringing cheers. Bill was prone to seasickness and every sailor not so inclined had goaded him all morning, trying to induce vomiting. They told him it was for his own good, get it out of your system and be done with it.

The rigid environment of a warship at sea sanctioned few distractions. Any break from routine was quickly transposed into entertainment, even another man's misery. Especially misery.

Earlier this morning, "Hey Bill, we're going to sea today," drained color from his face. While the boat was still moored fast, a voice outside Maneuvering called through the open door, "Did we just cast off? I think we did, didn't we? Aren't we underway?" Bill's knuckles whitened as he stared straight ahead at his electrical meters. The best performance of the dry heaves almost got the job done, but Bill rallied with a few deep breaths. Money was won and lost on how close to the hour he succumbed.

His suffering now complete, the trash bag lay twisted and tied at his feet. Men that teased him earlier now stopped by to make sure he was all right. Bill's leading petty officer stepped up to the chain across the Maneuvering doorway.

"Request permission to enter Maneuvering," called out the Electrical Division Leading Petty Officer, Chief Lethin. The Engineering Officer of the Watch responded with proper formality.

"State your business with Maneuvering."

Grinning over at the stricken Electrical Operator, the E-DIV Chief replied. "Sir, just making sure the Electrical Operator is going to survive the watch."

No equipment had gone wrong yet; the Engineering Officer of the Watch was still in a good mood. He volleyed the question with the formal informality common to submariners. "Electrical Operator, report the status of your gastronomical system."

Only now regaining color, the Greiner was too bleak to play. He fired back. "I'm here, goddammit. I'll stand my watch." Everyone smiled while keeping their attention focused on their work. The EOOW smiled at the E-LPO.

"There you have it, Chief."

The chief gave an "Aye aye, sir," snapped a salute and went back to monitoring machinery and watch standers.

The Electrical Operator did feel better now that the sloshing in his stomach had ended. A sympathetic soul brought wet paper towels and a cup of ice water. Greiner washed his face and rinsed his mouth. He asked a passerby to put the warm, heavy trash bag into a garbage can, preferably one out of smelling distance.

"Hell no!" the sailor responded, reeling away from the trash bag Greiner proffered. "Get that disgusting thing away from me." Sympathy has limits among men. Sympathy time was past.

The Officer of the Deck barked into the 1MC announcing system what everyone wanted to hear.

Dive, Dive, Dive. The klaxon blared, drawing out each raucous note; Aaooooooooogah, Aaooooooooogah, Aaooooooooogah. The song of the submariner.

Air trapped in the ballast tanks was released though the main ballast tank vents with a sharp whoosh. Seawater flooded the Main Ballast Tanks through grated openings at the bottom of the hull. Positive buoyancy lost, the boat began to sink. Bow and stern planes angled downward, wings in the water, guiding the boat down into the sea as her forward motion continued.

The bucking and rolling subsided as the cacophony of the surface faded. The dissonant complaints of the Main Engines eased into a steady heartbeat. Calm settled throughout the submarine. The wild activity of the surface transit shifted to a steady routine of men going about their business.

I marveled. I was traveling under water through the Pacific Ocean. Not on the ocean, in the ocean!

On your first trip out on a nuclear submarine, if you catch yourself staring at the overhead imagining the ocean above and the vastness below, don't be overly concerned. It's normal. It's exactly what I did.

6

Joining Up

The incentive for volunteer enlistment goes up considerably when being drafted is inevitable.

— *The Submarine Sailors' Imaginary Book of All Knowledge*

IT WAS SUMMER 1980. Teenage boys' and girls' hairstyles were much the same. Collar length or longer was hip, feathered back from an arrow-straight part was the look. Big Hair was in. Teen boys spent as much effort on their 'do as the girls. Arriving anywhere, the first stop was the nearest mirror for a quick touch-up. Fawning over your hair would get you branded as vain. Vain was not cool. Cool was looking perfect without trying. The trick to being correct—and still cool—was one or two quick comb strokes combined with a practiced head toss. Perfect.

The same low-rise bell-bottomed Levis we had worn with our platform shoes now dragged the ground over our blue Waffle Trainers made by a new company we mispronounced by leaving off the foreign long "e." We eventually learned to pronounce Nike.

Every male between the ages of 18 and 26 knew at least one friend with a poster of Farrah Fawcett. Her red swimsuit highlighted a wonderland that stoked the fire of our unrequited hormones. The real magic of that poster was Farrah's eyes. Caught in the same furtive glance we cast toward women, she looked at us the way we longed to be looked at.

The poets of our age were musicians. Bob Seger's "Against the Wind" lamented the frustration of our youthful ambitions. Supertramp's "The Logical Song" protested our involuntary domestication. Young men pining over a lost girlfriend sniffled through Michael Jackson's "She's Out of My Life." Every young person with a guitar could stumble through the first notes of "Stairway to Heaven." Cars full of teenagers rolled along singing about bad school teachers with Pink Floyd. The rumored breakup of the Eagles depressed us as the death of a close friend.

We went to the theater again and again to see *The Empire Strikes Back*. We thrilled to Luke Skywalker and lusted after Princess Leia as together they fought evil in the form of an aged Emperor and his overbearing minion, Darth Vader. Vader was the embodiment of the repressive authority we felt in our own lives. Do as you're told or be punished. We saw ourselves as young Luke, independent, courageous, and promising.

Our conversations were peppered with lines from the newest movie, *The Blues Brothers*. We mimicked Jake and Elwood flinching under the nun's yardstick. "We're on a mission from God."

Fifty-four Americans were hostages of political extremists in Iran. The term extremist was not in the vernacular of the time. We thought of them the same way we thought of airplane hijackers: thugs, bandits, robbers, and all-around jerk-offs. And where the hell was Iran?

We did not know, and did not want to know, much about the Iran crisis. It was a movie we didn't care to see. After all, it was a politicians' issue, like tuxedoed-and-gowned state dinners to which we would never be invited. The Iran hostage crisis was a news story from the other side of the world and therefore not real.

There was a religious element involved in the Iran hostage drama to which we did not relate. My friends and I had some basis in religion. Mesa, Arizona was a conspicuously religious community. But blindfolded hostages held at gunpoint on the evening news did not correlate with any religious principles we knew or could imagine. We had an equal understanding of Muslims as we did of Martians.

The Iran trouble confused me. I understood only two terms in the complicated equation. Americans were being held prisoner; and America had not rescued them. That bothered me. It went against the code. Never leave a man behind.

I had seen *The Deer Hunter*. I imagined American hostages forced to play Russian roulette for the amusement of their cruel Iranian captors. I knew that if I had been one of the hostages, I'd be mad that no one had come to get us. It was un-American. John Wayne would not have left those people there. He would have gotten them out himself. Unfortunately for the hostages, Big John died six months before that trouble began. U.S. President Jimmy Carter was no John Wayne.

To his credit, President Carter sanctioned a rescue attempt after six months of fruitless diplomacy. A military task force planned to sneak into Tehran in the night to kidnap the kidnapped. Now we're talking, I thought. Send in the Marines.

This valiant attempt turned into unmitigated disaster. Apparently the Armed Forces of the United States had been completely undone by desert sand.

I figured the people in charge of the rescue mission must be complete sissies, to abort the mission—some of our people killed in the process—because of a little sand in the breeze. Years later I came to understand the difference between the Arizona desert I knew, and the immense sandstorm that derailed Operation Eagle Claw.

The Russians had recently invaded Afghanistan, but my friends and I knew nothing of that. To us it was, once again, more grownup drama. Events like this gave adults reason to pore over newspapers and the stupendously boring evening news. They wrung their hands and complained about the government. None of that interested us. We were too busy hanging out at Jack-in-the-Box.

To display disapproval of the Soviet invasion, President Carter decreed that American athletes would not participate in the 1980

Summer Olympics in Moscow. We did have opinions on that issue. I, and everyone I knew, thought this decision was the very definition of bullshit. I was a track and field man, a sprinter, and the summer Olympics was a highlight of every fourth summer. At my age I could recall only the 1976 Olympics and had been anticipating the upcoming 1980 games. Now the Man had taken that away in a little-kid move. He took his Olympians and ran home in a huff. Never mind the athletes—his citizens made political pawns—who now had to sustain their peak condition for four more years.

Then the politics that I paid little attention to reached out of the television and grabbed me by the throat. On July 2nd, 1980 President Carter signed Proclamation 4771 requiring American males 18 to 26 years of age to register for the military draft. With a pen stroke the nightly news had become personal.

There may have been a valid military preparedness issue, I didn't know. In the blissful naiveté of youth, I by default entrusted the managing of mankind to the expertise and motives of those in authority. Perhaps the President was chest-pounding to impress the Soviets. Maybe beefing up the military would intimidate the Iranians who were clearly unafraid of our military might. Day 241 passed for the Americans still captive in Iran while the President was inking his name on what felt like my death warrant.

Whatever its purpose, Proclamation 4771 put a serious crimp on my plans. Prior to this, for my friends and me, debate and discussion about our futures had been somewhat of a dream sequence, a far-off fantasy where everything was possible. Now the future was here; the deck reshuffled in the middle of the hand. Another possibility had been added, but it was bleak.

If drafted, I would be powerless to refuse. I had neither the resources nor the inclination to dodge. If called, like it or not, I would go.

My buddies and I went to the Post Office and registered for the military draft at a time when our country seemed on the brink of war in the Middle East. Then we waited.

Whether or not a draft would actually be initiated was anyone's guess. The news media did big business with the issue. Experts expounded on the value of a draft and on the undesirability of draftees in the modern military. Experts discussed the likelihood of a draft and yet more experts explained how the military had no need of additional manpower.

There wasn't an expert available to tell me how to get on with my life with this ticking bomb of a military draft as a constant companion.

College was a possibility to avoid the draft; however, college was a problem. I had attended college two semesters. My academic achievements were unremarkable and I chose not to return for my second year. College was off the list. This left me with waiting to be drafted, or not waiting.

At some forks in a man's road the best direction is easily decided. This was not one of those. This was the most significant decision of my life. Should I wait and see whether or not the draft was revived, and thus go where conscripted, or should I preemptively make a choice of branch and join the military voluntarily?

If drafted I reckoned I would almost certainly become Army or Marine infantry. I would be a foot soldier, a dog-face, a grunt. I would be among the ones who always suffered the most and had the highest probability of coming home in a box.

I had seen Vietnam movies: The *Boys in Company C*, the *Deer Hunter*, and *Apocalypse Now.* I watched them as documentaries on my future as a draftee. The soldiers' life displayed in those films was unappealing. Walking to war wearing green and carrying a rifle seemed like the certain and horrible end of my life.

It wasn't the natural base fear of suffering, agony, and death that bothered me. Mine was a deeper, more loathsome nightmare. I was terrified of being disappointed with my life.

Armed Forces recruiters had come to Westwood High for Career Day. The Army spoke of its vaunted history. Its slogan,"Be all that you can be," I heard as diminutive, a declaration of the Peter Principle. The Air Force was all about flying, what else could they be, but airplanes didn't stir me. The Coast Guard was interesting, but not enough. The Marine Corps laid out its tradition of courage and honor, and its reputation for toughness. I thought Marines were a little too proud of dying. If I had to go to war, I wanted to be under brighter lights.

Then the Navy representative fronted the room. He was poised and appeared supremely competent. His dress blue uniform resembled a business suit with its white shirt and black tie. This uniform was less adorned than the Army or Marine Corps uniforms, with their pins and shoulder patches, but the Navy man had a self-assured presence that belied any tendency to measure him by colored ribbons.

The Army and Marines Corps recruiters swaggered, standing unnaturally straight, their chests puffed and arms cocked as if to throw a punch. They spoke with hard-ass, tough-guy voices.

The Navy man seemed amused by them. He quietly put up photos of amazing ships as complex as *Star Wars* battle cruisers. He spoke in an authoritative tone, like a respected teacher addressing the class. I leaned forward. He talked about airplanes and helicopters like the other services, but I ignored those topics. I liked the photos showing ships trailing a pure white wake through blue water. Cool blue water was a brilliant sales technique for kids in the Arizona desert.

The Navy man spoke of electronics and sophisticated machinery. He talked about things a warship did to remain independent. A warship. The term alone screamed adventure. Going boldly into harm's way, it was "Star Trek" come to life.

The Navy man told us that one modern nuclear powered aircraft carrier possessed more firepower and destructive capability than all the bombs dropped and all the artillery shells fired in WWII combined. Another masterful sales approach for young men drawn to horsepower.

The Navy guy used a magical word: Engineering. He used the word in every other sentence. He was a Navy engineer. I could be a Navy engineer. The Navy would train me.

He talked of submarines using nuclear fission to travel under the sea. Only the cream of the Navy was in submarines, he said. Every job was technical and demanding, each man tops in his field.

Atomic power. Splitting atoms. Fish on! I was hooked. If I had to go to war, I wanted to go with the smart guys.

Born early in February, I had always been among the youngest in my class. Skipping the 7th grade, I had graduated high school aged 17 years 3 months. My mother's consent was required for me to enlist at 17 and she would not give it.

My parents divorced when I was six. My father lived a couple of hours driving distance away. Once I had a car, he and I developed an arms-length relationship based on the shared experience of having lived with my mother. Hearing my military dilemma, he responded, "Whatever you decide will be fine with me." Supportive, yes, thank you—but not helpful.

My mother decreed against everything interesting. It's possible that she had said yes once, and regretting it, had conditioned herself to answer no by default. Yes was dangerous. No was safe. It seemed to me that no was her favorite word.

Mom offered no useful advice to help me determine what to do with my life. She just said no to everything that required her assistance or money. We were always short on money; my mother was shorter on assistance. She survived between alimony and child support checks. At various intervals she sold Avon or Fuller Brush or Shaklee products to fill the spaces between mortgage payments on our double-wide. It seemed to me that people bought products from her because they knew she needed the money. My birthday gifts from the kids next door were often Avon soap-on-a-rope which had been purchased from my mom. Often they were re-gifts of the birthday present I had given them. I still don't understand how anybody could think a ten-year-old boy wanted soap for his birthday.

My parents, running long on girls and short on boys, chose the efficient solution. I had been adopted at four-days-old. My mother, now divorced, fed me, clothed me, gave me shelter, and took me to church. That was the limit of services rendered. She was all in, tapped out. Her bucket was emptied. Worldly advice for a young man, more than a long list of don'ts, was beyond her. She could no more imagine a promising future for me than she could for herself. Her dictionary ended with "survive." The concept of making choices that might lead to a fulfilling and happy life was not in her repertoire.

My future undecided, my part-time job during high school became full-time. I worked as an air-conditioning repairman.

Then off I went to Texas for college, you've heard how that worked out. A college chum's family owned a dude ranch in Colorado, so I went there, wrangling horses and taking guests on rides into the high country. Moving on to California, I spent a few months repairing air-conditioning units and bar coolers at the Highlands Inn in Carmel. There was a girl there.

At the Highlands Inn I worked in the maintenance department with a dozen other men; carpenters, electricians, plumbers, and a TV repairman. One guy ran the sewage treatment plant. He ate lunch by himself.

These guys were all ex-Army who had been discharged at nearby Fort Ord at the end of their enlistments. They advised against anything military. "Don't do it. There's tons of bullshit, kid. Army, Navy, whatever. It's all full to the rim and overflowing."

Filtering the advice my ex-Army co-workers gave took ten seconds, even with my limited life experience. I wanted advice and took in all they offered. But these guys were Army; I was talking Navy. What did they know? More important in my decision making process, I judged them to be screw-ups, every one. They were California horticulturists—they grew their own smoking leaf—and they were drunkards. They all downed a little "bump" at lunchtime to get them through the afternoon. Their issues with the Army would not possibly apply to me. Chronic malcontents, I reasoned they were likely always in trouble. It would be natural for them to hate any institution that made their rebellious lives miserable. These guys started to twitch anytime a police officer was in sight.

The value of their advice I discounted heavily, pretty much down to worthless.

I, on the other hand, was not a troublemaker. I was a rule follower.

Way back in kindergarten, on an occasion when we formed a line to exit in an orderly fashion, I screwed around as five-year-old boys will. The teacher, Mrs. Taft, thumped me on the head with a bony finger so hard that my head rang.

"Behave," she said. "Stand still."

Mrs. Taft was my next-door neighbor and was always nice to me. Beyond the pain in my head, the shame of disappointing her was a hot flame. From then on I stood quiet in line. I did as I was told. I became a fast learner. Shame does that for a boy—or for a man. The Navy would have no trouble from me. I knew how to behave. I would be fine.

One issue remained that gave me pause. The Navy required a six year enlistment for the advanced field I wanted. That was contrary to a lifetime of adults advising me to be careful what I signed. How did I know if I would like it or not? Shouldn't there be a trial period, a test drive, or an apprenticeship? Once I signed, I was in for the duration. The Navy had a "no returns, no money back" policy. All sales were final.

There was also a nasty caveat in the agreement. If I didn't keep my grades up, or failed any of my schools, I would be dis-enrolled from the Nuclear Power Program. I would still be in the Navy; I would be U.S. property to be disposed of as they chose. Failing out of Nuclear Power I would be at the mercy of the "needs of the Navy." All military and ex-military guys I knew told me that if I bombed out of school, my next assignment would be a place in Egypt with a vulgar name. Now how could they know that?

Of all the pros and cons I wrestled with, there remained one pro I could not deny, nor argue against.

Adventure.

One of the main reasons I had left my hometown in the first place was to find a life different from the one I would have there. Draft or no draft, war or no war, if I joined the Navy I would never sit on the porch in my dotage and feebly wonder "what if?"

I motorcycled into Salinas and enlisted in the Navy to become a submarine nuclear reactor operator.

7

Ten Weeks

Most people never find out what their one-hundred percent looks like.

— *The Submarine Sailors' Imaginary Book of All Knowledge*

THE 1987 STANLEY KUBRICK film *Full Metal Jacket* is the quintessential Basic Training educational piece. Although set in the Marine Corps; the movie causes all military personnel to chuckle and groan recalling similar experiences.

The premise of Basic Training is simple. Convert coddled undisciplined civilians into self-sufficient military members in ten weeks time. It is a simple idea, but a complicated task. If you are reading this hoping to get some new information that will smooth the way of your upcoming Basic Training, sorry. I've got precious little advice for you. What I will give you follows.

Do what you're told and don't fight it. The professionals you think are sadistic bastards have been hard-trained and their competencies evaluated. They know what they are doing. Basic Training has been around for 7,000 years, since the beginning of war. Basic Training is not a random exercise. Later you will come to know that the Navy has boatloads of people who monitor training effectiveness and adjust accordingly. Millions of American fighting men have gone through Basic Training before you. All learning requires discomfort. You will survive, quite nicely in fact.

At Naval Recruit Training Center San Diego, eighty men formed a Company. Originating from every corner of the country, some guys were 17 years old, most 18 to 20, with a few "old guys" at the advanced age of 23. They came from every conceivable background and social strata. Some had parents who were doctors and lawyers, PhDs and scientists. Some had existed on Welfare. Some had never met their parents.

Every major religious belief was present, along with a few denominations unfamiliar to me. I couldn't help but stare at the men speaking hard New Jersey lingo as well as the Deep South boys. Every imaginable hairstyle and attire had a representative.

There were the hippies, the stoners, tough guys and cowboys, the street-wise and the rubes, the preachers, the scholars. Smart guys and dumbshits. Good guys and assholes. The IQs present ran from minimum acceptable to Mensans. All recruits had graduated high school or its equivalent. Many had a year or two of college. In my boot camp company, several men held bachelors' degrees. Even though they could have applied for an officer program, some of these guys were compelled by family tradition to serve enlisted. Many of those planned to enroll in an officer program once in on the ground floor, so to speak.

In hindsight, that strategy was unsound. One of the smartest men I ever knew entered the Navy enlisted, planning to enroll in a commissioning program. He enlisted with a Bachelors degree. He completed a Masters degree in psychology at the University of Hartford while he and I were students at the S1C submarine prototype in Windsor Locks, Connecticut. He was working 12-hour shifts 21 days a month. To become commissioned as a Navy Ensign—with a Master's degree and 4.0 evaluations—took him six years. Not exactly the fast track he was hoping for.

Bureaucracy is geared for maintaining the status quo, not for advancing individuals outside the norm. Abilities and qualifications do not trump the system.

Looking back on Boot Camp with the luxury of retrospect, the Navy does a masterful job. In less time than a school semester, young men are transformed. They enter Basic Training as random individuals and leave as Navy men. They leave with a heightened sense of self and a new-found confidence. Regardless of who or what they were when they started, they went out better men.

While many improvements did occur, Basic Training was not designed to be a personal growth experience. It's not an EST seminar.

The opening scene in *Full Metal Jacket* portrays Gunnery Sergeant Hartman explaining to the new recruits the boot camp program with wonderful precision. Although the language is coarse enough to give a doting mother an aneurysm, the scene is beautiful in its impact. In one two-minute monologue, the Senior Drill Instructor accomplishes several things. He asserts himself as the Man-in-Charge. He defines his purpose for the recruits. He takes away their illusions; he smashes the world they have become comfortable in and introduces them into a new world. The new world doesn't care about hair style, what car they drive, or in which part of town they live. This new world is filled with performance standards and self discipline. The Senior Drill Instructor will repeatedly challenge them to do better.

Their new world is one where accountability is a higher priority than in the world they left. No longer can they get away with schoolboy answers such as, "I forgot." The Drill Instructor—Company Commander in the Navy—disturbs the recruits and makes them uncomfortable, causing them to pay close attention. Rebels in the company will single themselves out for special discomforts, while the others tremble in their stiff new boots hoping the Senior Drill Instructor does not look their way.

In *Full Metal Jacket*, although Sergeant Hartman yelled at the men, there was no anger in it—for the most part. He spoke assertively, yes, and loudly, his words bludgeoning their way inside the man's head. The Senior Drill Instructor spoke to the center of the man's brain, no hemming and hawing around the edges. The volume and heat of the diatribe disconcerted the direct recipient, but also filled the space. He had no public address system and every word he spoke, he spoke to 80 men. The mistake of one man becomes a teachable moment for all. Eighty men learn from one man's error.

When Gunnery Sergeant Hartman is eyeball to eyeball with a recruit, the Gunny does not blink. That unnerves a man. The recruits must learn to operate under pressure, to stand and deliver, even when they would rather be anywhere else. The world they are entering makes no allowances for slow learners.

My recollection of the first day in Basic Training is of how long it went on. I kept feeling that at some point I would be done for the day. An eight-hour workday had passed eight hours ago and we were still at it.

"That's all for today, kids, we'll see you tomorrow," were hoped for words that never came. There was no other place I was supposed to be: No ball practice, no after school job, no homework to do. My mother was not expecting me home tonight.

Bankers' hours of nine to five were never to be seen for the duration of my Navy career. In boot camp we rolled out of our bunks at five in the morning and did not return to bed until eleven or twelve o'clock at night if we were lucky. The daily schedule was sacred. Sleep was optional. If we fell behind, we didn't sleep until we caught up. There was no, "It's late, we'll finish tomorrow." We finished the task in front of us, every time. Tomorrow there were new things to learn.

Our rising in the morning was not leisurely; it's amazing what a man can do in 15 minutes when motivated. By the end of the first day I had a new hairstyle and a new wardrobe with my name and Social Security number stenciled on every item. I know it was stenciled on every item because I put each stencil in the place instructed. If a man made a mistake with the stenciling pen, he lined-out the mistake and corrected it. He wore his mistake for the next ten weeks.

Punishment? Perhaps. It was also a life lesson for all who saw the man's lined-out and corrected name blaring on his shirt like a "kick me" sign. Some mistakes last. This was only day-one; the lessons were just beginning.

Boot Camp was jammed with medical exams and inoculations. We marched to classes about Navy organization and history, military ranks and etiquette, first aid, advancement opportunities, and how to avoid venereal disease. We experienced fighting real fires, tear gas, and continuing to operate while sleep deprived. The instructors pushed every button we had and invented a few of their own.

I well recall the shock of grownups in authority cursing. And not just cursing, but vehement, violent, over the line stuff that anywhere else would have gotten them punched in the face or arrested. I had worked in construction zones and was familiar with rough talk. Most of the boys I'd grown up with used coarse language when they could.

I'd experienced harsh words in high school sports, but in high school that smacked of obvious criminality. Those coaches would have spoken differently if a parent or the school principle had been present. That was low-grade tough guy stuff, like jaywalking only after you've made sure a police officer was not watching.

The chief petty officers screaming obscenities at me now were professionals who took pride in their work. It disconcerted me.

I was raised under the weight of a devoutly religious mother; hearing "goddamn" screamed over and over caused me to wince skyward expecting the apocalypse. It never came and I got over any squeamishness with mere words. Goddamn right I did.

We marched—in ranks and in step—as a company everywhere we went. We ran for miles in ranks and in step. For months after Basic I found myself counting cadence as I walked down the street in civilian clothes.

Close order drill was rife with teachable moments. One man turning left when 79 men turned right is not an error that goes unnoticed. How many lessons could be hammered home in that one instance?

The world did not stop turning because a recruit had a blister or a hangnail. Being tired would have likely resulted in special treatment at home. "Now, now, dear, lie down and rest. I'll call you for supper." Not here. As P.T. Barnum well knew; one monkey does not stop the show.

Locker inspections and bunk inspections were also fertile ground for teachable moments. The United States Navy does not care how you fold and store your socks once you leave Recruit Training. They do care that you can understand and follow procedure. No detail, regardless of how minor or trivial it may seem, must escape you. That's the lesson.

One Boot Camp memory is with me still. Basic Training was nearly complete and we were undergoing a locker inspection. After ten weeks, the only thing I knew about the Company Commander was that he was a Navy Nuke, the same career field I wanted. He wore the silver dolphin insignia of a submariner. He was also a squared-away guy. He exuded composed competence. His example convinced me that I was on the right path in seeking to become a Nuke.

Finding my locker to be SAT—the unpleasant alternative being UNSAT—the Company Commander looked at me as though for the first time. He peered at my name stenciled above my uniform pocket.

"Heckman."

"Yes sir." He straightened and looked me square in the eyes. He was a tall, bearded, dark-haired man with pale blue eyes. His appearance was striking. Those eyes contemplated me for a long moment.

"Do you do everything with the same attention to detail you give to this locker?"

My past could not justify such a claim, but I knew that in my future there could be only one answer. "Yes sir."

"Then you'll make a fine Navy Nuke." He moved on, leaving me in stunned amazement. I hoped he was right because at this point I was not feeling particularly secure in my new career.

I left the ten weeks of Basic Training with a clear understanding of accountability, responsibility, and attention to detail. I had seen what those attributes looked like in real life. I knew what verbatim meant and understood zero tolerance for mistakes. In my new life a mistake at the wrong time may result in more than a minor inconvenience like a poor mark on an exam. One mistake could cause death and destruction, perhaps my own.

Regardless of service affiliation, all Boot Camp survivors have stories of the experience. If your grandpa, father, uncle, brother, hasn't shared those stories, prod them. As years pass, many stories expand in scope and hilarity, as do their tellers.

Old soldiers and sailors feel no shame in telling of events and situations that occurred for someone else, yet begin those stories with "I was there when…"

Sometimes the truth is closer to "I knew a guy who had a friend who told him…" In either case, the factual origin of the tale is long forgotten and the memory is now theirs. The following story I personally witnessed. As Navy men say; this is no shit…

Basic Training nearly finished, we were given our first liberty. The Navy is replete with deceptive terms. For an infraction you are "awarded" a punishment. A reward for exemplary service is a "citation." However, "Liberty," is a traditional term that even in the Navy means just what Mr. Webster had in mind. Walking out the front gate of Recruit Training Center for the first time, not having to be back for 48 hours was a magnificent feeling. A man walking out of prison must feel much the same.

Outfitted in crisp new dress blue uniforms—Cracker Jacks by nickname—Seaman Recruits flew to alcohol and tattoo parlors like bees to the flower. They sought the former first and, once enough courage had gone down their throats, acquired the tattoo they had been obsessing over. For the past ten weeks, at every break given to shine shoes, write letters home, and "smoke 'em if you got 'em," conversations centered around tattoos.

"Man, you show up in the fleet without ink and they know you're just a baby," said a recruit who had been discharged from four years in the Marines and was now going Navy. His arms and shoulders carried slogans. "Kill them all, let God sort them out," "Death before dishonor," and "Semper fi." After I asked if quitting the

Marines for the Navy dishonored the Corps, he glared at me with what I believed was his game face for killing.

As for first impressions, I figured the lack of chevrons and service ribbons would signal my inexperience louder than any tattoo could overcome.

Returning from liberty late that night, the barracks reeked. The tea-totalers trying to rest on the peaceful laurels of sobriety were bludgeoned by the boozy exhaust of the blissfully passed-out.

"God day-am," whispered one of my companions when we came through the door. "None of you morons light a cigarette in here or we'll be in tomorrow's newspaper." Worn out as we were the absurdity refreshed us and started us sniggering and making up headlines.

"Barracks Blows Sky High."

"Navy Tests Secret Explosive."

"Raining Seamen in San Diego." I nearly strangled trying to hold in laughter. Our fun was hushed by the Fire Watch. He shined the duty flashlight in our faces.

Through clenched teeth he said, "You guys shut the fuck up."

The next day a severely hung-over recruit gingerly peeled the bandage off a fresh tattoo. On his arm was a bright red heart bisected by a blue banner. The skin under the tattoo, and the area surrounding, was an angry pink color. An onlooker cocked his head to make out the scripted banner.

"Becky," he read aloud.

"It says Betsy," said the proud owner. "She's my wife."

"Sorry, recruit, I don't care who you're married to, but this here tattoo says Becky." The man jabbed his finger into the word, causing the owner to curse and pull away.

Damn. That Jose Cuervo was no friend of his. The men crowded around, guffawed and shot tough-guy condolences toward the unhappy man. "Dipshit." "Moron." "Dumb motherfucker." Those with new tattoos sneaked off to the shaving mirrors to more carefully inspect under their own bandages.

"Goddammit. I'll never go back to that tattoo joint again," snarled Becky's Husband. But of course, he went back that night.

The tattooist who did the original work was not to be found. It seemed the tattoo shop had never heard of him. Perhaps the man's name spoken in English—Kentucky-style English—was the real mystery to the proprietor. The tattooist on duty knew just enough English to repeat over and over, "No money back."

"See. You see," he said sternly, jabbing a finger at a hand-lettered sign thumb-tacked to the wall:

NO MONY BACK

Unable to erase the error, the incriminating name was buried under red roses and green leaves while our man ground his teeth during the procedure. Next morning, word of the kerfuffle reached our company commander.

He stepped up to Becky's Husband and made as if to punch him on the affected arm. Becky's Husband steeled himself for the blow that never came.

"I hope you asked for the clean needle, the one he uses on his mother," the Company Commander said. Inspecting the affected area he rendered his opinion, more bemused than concerned: "that arm just might rot off." Becky's Husband did look feverish.

It was the only time I saw our Company Commander laugh.

After boot camp I felt less in common with friends back home. They were all fine people, we had great history, but they now seemed sloppy and disorganized. This generalization was unfair, but I felt I had faced the monster and lived to tell. My old friends had stayed home and let the tide of cable television wash over them each evening. Every day they were softer than the day before.

No longer the scrawny teenager who left to join the Navy, I felt I was now a man, proven and hard. But at only 19 years old; the only thing I'd done was graduate recruit training. Me and fifty million other Americans before me. What the hell did I know?

I did carry new confidence. Not cockiness, not arrogance, but a deep assuredness that whatever challenge the day would bring, I was up to it. I was a Navy man.

8

School Days

The government educates you and trains you, and then gives you a fine position for life, for which, in return, you agree to do whatever the government demands.

— *The Bluejackets' Manual, 1940 edition*

GRADUATING BOOT CAMP I boarded a Greyhound bus headed for Salinas, California for my first paid vacation ever. You'll recall there was a girl there. Anticipating our reunion had given me refuge through Basic Training. Turns out that she was far less thrilled to see me than I her. In the three months I had been away she had moved on with her life. I don't know that we were ever a serious item, but I recall disappointment settling deep inside me at what felt like the sudden change in her affections. We were strangers. Her shock at seeing my obviously military haircut registered clearly on her face and confused me. I think she planned on never seeing me again yet here I was on her doorstep.

I'm not sure that I've ever recovered from that slow sinking sensation of realizing that I had overestimated my value. A therapist could probably tell me how that early event shows up in my life to this day.

She tried to be supportive, bless her, but her heart wasn't in it. "What are you going to do now?" she said, sounding as though my next move was up to me.

"I'm going to school to be an Electronics Technician," I said.

"Where's the school?"

"The first one is in San Diego."

"Oh…" she said. And that was that. The next day I boarded the same Greyhound now headed back to San Diego. I've not heard from her since.

As you surely have surmised, training is required before being permitted to operate the reactor on a nuclear submarine. In fact, quite a lot of training. Prior to reporting to my first submarine, I attended six schools in four states over a period of two and a half years.

An additional remedial course was required of me in recognition of my unremarkable math skills. Scoring the minimum allowable passing grade on the Nuclear Field Qualification Test, I attended mathematics refresher training prior to Nuclear Power School. Ten percent of Nuclear Power School class 8203 joined me there, gaining the algebra, geometry, and calculus prowess necessary to survive Nuclear Power School.

To maintain my position as trainee in the Nuclear Power Program, passing each course was not enough. My grade point average was required to remain in the upper twenty percent of every class. If my final class standing fell below that mark, I would be dis-invited from the Navy's Nuclear Power Program and sent to the fleet to make myself useful. "Haze Gray and Underway" I would be. I would be getting plenty of sun while the pasty-faced "Upper Twenty-Percent" still in the program would long for the Florida sunshine from the windowless confines of Naval Nuclear Power School Orlando.

The Navy has been at the business of technical training long enough to have a firm handle on the effort. Each man was trained in a non-nuclear field and had to prove his ability to learn before advancing to nuclear training. If a man failed to maintain the standards of Naval Nuclear Power, he could still be a productive member of the Navy in the field in which he had been successfully trained. In either case, the Navy won.

Nuclear operators must have comprehensive knowledge of a long list of machinery and engineering processes. Schooling a Nuke took a couple of years and that "couple of years" was unlike high school or college. In Navy schools we went to class all day, five days a week with duty days—a work day, usually barracks watch—on weekends. Summer vacation? No. Spring Break? No. A month off for Christmas? Again, no.

Failing a course in college, civilians get a do-over, although they will pay for the privilege. Failing a course in the Navy would have me finishing my enlistment doing something other than that to which I aspired.

Reactor Operators were first trained as Electronics Technicians since much of the equipment we would be responsible for was electronic reactor control systems. Charged up to prove my worth, I hustled through the self-paced modules in Basic Electricity and Electronics school in San Diego. This notion of meritocracy was a mistake. I now wished I'd paced myself in "B double E" for a marathon rather than a sprint. I thought my hard work would be noted and parlay into advantage. But in any bureaucracy as ponderous as the United States Navy Training Command, a protocol for merit based advantages does not exist. The training system is a rigid pipeline. Nothing speeds or hinders the flow.

Great Lakes Naval Training Center was better known as "Great Mistakes" by sailors bemoaning the region's weather and schooling in general. I didn't care about its reputation; traveling to a new place was an adventure. I had never been east of the Mississippi. "Join the Navy and See the World," claimed the recruiting poster. I was being paid and the travel was free. I loved this Navy thing.

Out the door of Chicago's O'Hare airport I was welcomed by a stiff March breeze under a gray sky heavy with moisture out of Lake Michigan. My new love affair with travel was temporarily suspended. My short-sleeved white uniform had been perfect for sunny San Diego, but the Chicago wind was not deflected by the thin cotton and chilled me to the core. The viciousness of the cold unnerved me. What had I ever done to Illinois? Great Mistakes indeed. I hated the place already.

Great Lakes Naval Training Center was a massive enterprise. I don't recall the entire list for 1981 when I was there, but the current list of schools located there will give an idea of scope. Fire Controlman (FC), Gunner's Mate (GM), Interior Communications Electrician (IC), Boatswain's Mate (BM), Hospital Corpsman (HM), Electrician's Mate (EM), Culinary Specialist (CS), Operations Specialist (OS), Hull Maintenance Technician (HT), Damage Controlman (DC), Engineman (EN), Gas Turbine System Technician (Electric) (GSE), Gas Turbine System Technician (Mechanical) (GSM), Machinery Repairman (MR), Quartermaster (QM), Machinist Mate (MM), and my rate, Electronics Technician (ET).

"Snipes Castle" was the massive barracks for Enginemen, Boiler Technicians, Electricians, Hull Techs, and others. The bachelor enlisted quarters for electronics rates, known as "Twidget Hall," was my home for nearly a year.

Uniforms of other services and nationalities attending school as official guests of the Navy were everywhere, splashes of color painted on a navy blue background. Coast Guard, Marines, and Air Force were all present as well as Canadians, Saudi's and a few uniforms the nationality of which I never knew.

At 0700 we fell in for morning muster on a grassy slope near the barracks. One morning, waiting for the senior student-in-charge, we watched a car being parked in the lot in front of us. Parking parallel between two cars, the driver alternated between Drive and Reverse, each time ramming the car in front and behind solidly. Forward, Bang. Reverse, Bang. Forward, Bang.

"Son-of-a-bitch needs an ass whipping," a terse voice growled from the ranks. Speaking in ranks was forbidden, even when just waiting around. As with all things forbidden, it was done discreetly. The trick was to remain motionless at "parade rest," keep your eyes forward, and talk quietly through your teeth, moving your mouth as little as possible to avoid drawing attention and retribution.

"There's no excuse for that bullshit." A murmur of agreement rustled through the company.

"Probably a fighter pilot," someone said dryly. A snicker rippled through the ranks.

"Definitely officer material." This time-honored enlisted joke fit all displays of mechanical ineptitude. It induced another titter.

"If he's training to be a tank driver he'll do fine." Shoulders quivered as a stifled chuckle ran though us.

"Hey!" a voice cried, startled by the recognition. "That's my car he's banging into!" Those last few holdouts who had been restraining themselves with proper military formality, at last blew their cover. The entire company shook with laughter, faces tipped back to the sky, mouths wide with mirth.

Finally parked, the car door opened and a gold-embroidered Saudi uniform climbed out. The company quieted.

"Mu-ther-fucker," drawled a disgusted voice. The car's driver walked up the hill and joined the Saudis grouped separately from our ranks. We grit our teeth. Welcome to America.

We didn't want to be prejudiced but—Christ on a crutch—learn to park a goddamn car. How about a little respect for the property of your fellow man and your host country?

I had been concerned about the U.S. selling war materiel to foreign countries. It seemed crazy to arm another country with our stuff. We were only students, but we figured we were military men and therefore, insiders. This, in hindsight, was comical. We were professional students talking without any knowledge whatsoever of the real world of the military. The apex of our careers thus far had been finding the chow hall the second time.

Still, selling arms that might one day be pointed back at us seemed foolish. After going to school with the Saudis, my position on the matter changed. I doubted that any warplane or ship we sold was still functioning one-hundred percent. I imagined Saudi fighter pilots using flashlights to see the working half of their instruments.

I'm not saying there aren't smart Saudis. I'm saying the gents I went to school with at Great Lakes were not going to trouble-shoot and properly repair much of anything. I did not know their culture or their boyhoods. Maybe they hadn't taken apart toasters, adjusted bike chains, or tuned-up their mom's car enough to gain a basic understanding of things mechanical.

9

Busy Work

No Commander shall inflict any punishments upon a seaman beyond twelve lashes on his bare back with a cat-o'-nine-tails...

— *Rules for the Regulation of the Navy of the United Colonies, 1775*

IN THE BOWELS of a Navy administrative office somewhere is a smoke-filled room. Detailers and school coordinators, woozy from too many whiskey sours the night before, plot the Navy man's future as a manufacturing project.

One school's graduation date does not necessarily coincide with the following school's start date. The schools are at different facilities across the country. Time must be allotted in the schedule for travel and possible leave. Cushion is built into the system. From the scheduler's viewpoint, it is better for a trainee to wait for weeks or months than to have one hundred students show up at a school with a ninety-student capacity. Empty seats would be a problem as well.

In a bureaucracy, nearly every inconvenience can be attributed to somebody covering his ass. The cumulative effect of many somebodys covering their individual asses is that the scheduling cushion grows. Between schools we waited.

There was no place in military tradition for enlisted men to simply go home and watch TV until their next school was ready for them. One is on the payroll and will earn one's keep. A temporary job was assigned, an in-between-schools gig. The job was not so much busy work as it was a menial task that could be done by any body still breathing. Sometimes the temp job was a welcome change.

Arriving at Great Lakes early for Electronics Technician "A" School and assigned to Grounds Maintenance, I mowed lawns. Sunshine, blue sky and spring breezes made the job pleasant. Timing being everything, the same job would have been miserable or impossible in December.

Recently advanced to Petty Officer Third Class, E-4, I was senior to many students waiting for their schools. When there was one job left on the list, it landed on E-2s and E-3s. I was set free one lawn earlier than they. After all, rank has its privileges.

On finishing ET "A" School five months later, I waited several weeks before Radar school. This time I was assigned to an inside, clean-hands job. I worked for an office of civilian ladies who maintained service records for the students. My job was to place incoming security clearance information into the appropriate service members file.

A service record begins when a person enters the military and follows them throughout their career. On exiting the service, that "Service Jacket" is warehoused until doomsday. Only in Hollywood's imagination does a service record from twenty years previous appear within minutes of a character saying, "Who is this guy?" Another character says, "I made a call. Look, here is his service record." Dream on, TV and movie people.

The ladies were fun to work with. So far, most of my Navy career had been spent in tightly controlled circumstances where my opinion was not only unneeded, it was unwanted. Go here, go there, do this, do that. It was a nice break to work in a free breathing environment.

A messenger, Leon, came and went several times a day delivering crates of documents into and out of the office. His car was a metallic green, 70s-era Cadillac, which he drove with style. Slow and easy, baby, Leon was never in a hurry. The car had every customizing feature that the auto parts store had on the shelf. Curb warning indicators, racks for cassette tapes, a stick-on thermometer, and green fur on the dash. If the end of a workday coincided with one of Leon's runs, he would give me a lift to my barracks. During the 10-minute drive to the far end of the base where I lived, Leon's car stereo played Soul music and Rhythm & Blues.

Marvin Gaye I knew, but now I met Al Green, Teddy Pendergrass, Bobby Womack, Sly and the Family Stone, and others. I think Leon felt it was his duty to educate this Wonder-Bread, pop-music white boy into the ways of smooth. Thank you, Leon.

The ladies in the office were mostly matronly and seem to enjoy having a young man around to tease. There was always an extra cookie or homemade brownie in their lunch. They also may have had a secret contest to embarrass me. A sign on the office wall read:

SEXUAL HARASSMENT WILL NOT BE TOLERATED
HOWEVER
IT WILL BE GRADED

The students whose files I kept would ultimately have different levels of security clearance depending on the job they were studying. Regardless of the clearance level, the first steps for all clearances were the same.

A notice was sent to the police station of every town in which the service member had lived. The form stated very politely that the above named individual was applying for a security clearance with the U.S. Navy, and will they please report all crimes the above named individual has ever been charged with? Please indicate convictions.

How easy was that? Another office somewhere did the addressing and mailing. I filed the returned completed form.

Out they went, light-green postcards addressed to cities and towns across America. Out to the big cities they went: Omaha, Dallas, San Francisco and Los Angeles, Seattle, Miami and Philadelphia, Buffalo, Richmond and Denver.

Out these inquiries went to small towns: Belgrade, Montana; Paris, Tennessee; Rome, New York; Athens, Georgia; Moscow, Idaho; Jerusalem, Utah; London, Arkansas; Dublin, California; Warsaw, Kentucky; Birmingham, Michigan and Cambridge, Nebraska.

Into the office they came, these same postcards. Each one was returned stamped by the authorities on the other end. The preprinted stamped impressions were different color inks and different fonts. Some stamps were neatly inked along the bottom of the post card. Some cards were stamped indiscriminately across the printed request. Some were all capital letters.

All the stamps read the same: "The requested information is releasable only to Law Enforcement Agencies."

The U.S. Navy was not a law enforcement agency, therefore, the form was returned blank. No information was gathered, nothing learned. It was a futile exercise. But the form was properly filed in the appropriate service record. Years later I saw in my own service jacket that same stamp across that useless form, carefully documented and kept.

We knew what the response would be before the card was mailed. I mentioned this revelation to the ladies in the office. Surely there is a better way to do this, I suggested. The ladies looked at me with long-suffering patience.

"Hush now, you're talking me out of a job."

No, ma'am. I'm suggesting that this can be done better and there must be a way to make it useful. I gave a few suggestions as to how that could happen. The ladies sighed, clucked, and went on with their endless filing. Such a child, they must have thought. He doesn't know how a government job works.

On we went; creating, mailing, and filing documents with no information on them.

The Petty Officer in Charge of finding busy work for in-between students chose me for a new job opening with Shore Patrol.

Cool, I said.

Saying goodbye to the ladies in the filing office, I hustled to the Master-At-Arms, Great Lakes Naval Training Center. The quarterdeck of the Training Center's police station was spotless and very impressive, with polished cannon shells and other shining Navy accoutrements. Over time, I disparaged spotlessly clean and shiny spaces. They don't get amazingly clean and shiny by accident.

The Navy does not have a janitorial staff. The Navy has sailors, and sailors can do anything. They do everything. Mopping, waxing, buffing the deck and polishing the bright-work were as much a part of being a sailor as was griping. In Navy buildings the floors were stripped, waxed, and buffed every night. I know this as a fact because I did my share. The Great Lakes Naval Training Center Master-At-Arms Quarterdeck was one of these.

However, this shiny deck—the floor to landlubbers—was not my primary duty. I was to be Shore Patrol.

Before going out into the public with authority on my team, some learning was in order. While Shore Patrol, name tags and service ribbons were not to be worn with our winter working blues. Our neckties were to be clip-ons. What the heck for? I asked. I despised clip-on ties as the ultimate in fake, right up there with a dickie.

My question was answered. These things were done to avoid giving an unruly sailor a handhold on your person. Especially grim could be the outcome of a preinstalled noose in the form of a necktie, its free end hanging down my middle. The precautions sounded like great ideas.

A grizzled Master Chief with tattoos of naked ladies in green on his forearms instructed us new guys in the art and science of the nightstick. It was a short class. He spoke with enthusiasm and with a disturbing gleam in his eye.

"First, you tell 'em to 'assume the position' against the wall," he said. "The assholes you are going to be dealing with are familiar with it, I promise you. Then you spread their legs out, wider, like so. Keep one hand on the back of his head. If he gives you trouble, you kick his leg out—like this—and with your hand on the back of his head introduce him to Mr. Wall."

He proceeded, teaching us the most effective way to bring a man down with the baton. If he is coming at you, jam the end into his middle. If there are two of them, whack here, then there.

Was I in the wrong class? Did he realize we are not trained law enforcement types? We were students in Electronics and Electricity with a few Enginemen and Interior Communication guys mixed in. We were twidgets and geeks. The Man was teaching us how to beat others into submission without leaving marks. Holy shit, Batman...

Apparently it was all on the up and up. After a half hour of nightstick training, and another half hour on wrongdoing to look for, we were on our own. The Master Chief signaled our Shore Patrol school graduation with this admonition: "Now go forth, and wreak havoc on your fellow man."

In pairs we went into the world wearing armbands with SP on them. We wore white canvas belts with our trusty baton riding in a steel ring. I practiced twirling the nightstick on its lanyard. Two minutes later, my forearms bruised and a lump rising on my forehead, I gave up nightstick twirling.

Being curious and sometimes stupidly so, as a test to see what it felt like, I tapped my shin with the nightstick employing a minor stroke, the tiniest of thumps. The pain sucked the air out of my lungs. I tried to imagine receiving a full stroke on my shin, one of the Master Chief's favorite tricks of the trade.

"Only use that when the situation has escalated to where the issue of leaving a mark is no longer a problem," he had said. Right. He seemed not at all concerned that we might actually find ourselves in such an altercation. I noted that he did not instruct us in how to extract ourselves from such, or how to get reinforcements. It seemed plain he expected we would whack our way to success and drag the culprits, all the culprits, in by the collar.

"Excuse me, Master Chief, but what if we end up with a crowd on us?"

"Get on the radio," he said. "Call me."

One riot; one Master Chief.

For the most part Shore Patrol was another Navy duty that was mostly uneventful boredom. My partner and I strolled along our assigned route. It was a pleasant walk. The idea was to stay uninvolved until we were compelled to get involved. It wasn't our job to provoke. People went about their business around us.

Occasionally someone called out, "Howdy, Shore Patrol!" We smiled and waved. That was one to watch, thinking he was clever by initiating preemptive goodwill.

Occasionally a jokester feigned terror. "Please don't beat me with that stick, Mr. Shore Patrol!" I always wondered if the man had encountered the business end of the Master Chief Master-At-Arms.

That was our public interactions during the evening watch. Then we drew the midwatch. No more strolls in the gloaming when social drinkers were enjoying a little buzz. Now we were on the job during the witching hours, when fun-loving easy-going drinking men transformed into bona fide shitheads.

Our average patrol out in town along the Strip, the row of bars just outside the base, would go something like this: There he was, the staggering drunk, thrown out of the bar and incensed over the injustice. After all, he has been drunker, lots of times, without being cut off.

We Shore Patrol, we agents of law and order, we servants of the greater community, we show up and say: "Hold on there, sailor, you look a little unsteady. Can we give you a hand?"

"Fuck you!" he screams in our faces. Ladies and gentleman, tonight's floor show begins. First drama, then a rodeo.

A radio call followed by the arrival of the paddy wagon. A white Navy van with two benches in the back bolted to the sidewalls. A steel grate separating driver from passengers. The floor is bare metal making it easy to hose out.

The Master-At-Arms guys that jump out are actual trained law enforcement types. We are their early warning system. We go out and find trouble, then call them. Proper wielding of our nightsticks should keep us alive until they arrive.

They take down the swinging drunk in a flash. He is handcuffed in the back of the van before his pals can come to his rescue. That would be a poor choice on their part anyway, but drunks are poor decision makers. The Master-At-Arms men don't give an uprising time to gain footing. They are a strike force, in and gone, the white van disappearing around the corner. The drunken sailor is whisked off to the brig and we continue our stroll. Shore Patrol at Great Lakes was an efficient business.

One Sunday night in January my partner and I were having an uneventful shift. At 2:30 in the morning we heard a faint rumble in the deserted streets of the base near the Bachelor Enlisted Quarters, the BEQ. If there was to be trouble, it would be there. The rumble grew into a racket, and the racket grew steadily until cresting a rise in the road, a very drunk man was rolling an empty aluminum beer keg toward us.

Pavement in the Chicago region sported basketball-sized holes left by chunks of paving material ripped out by winter ice. Lumps of asphalt patched holes left by winters past. This moonscape gave the keg roller fits. He tried to lean on the keg for support and roll it at the same time which proved far too challenging. His balance definitely not cat-like, his massive level of inebriation would have laid out a tent full of Cirque du Soleil acrobats. Each time he fell, the empty keg rolled away.

He cursed quietly, weary of the epic undertaking. "God-dammit, son'fabish. Git back 'ere."

The din of the empty aluminum keg scraping and bouncing over the pavement in the late-night, early-morning quiet was nerve shattering. A window opened in the BEQ and a "Shut the Hell Up!" roared out into the darkness.

My partner groaned. "Oh, shit. A quiet night shot to hell."

We stopped Keg Man. On this 25-degree night my partner and I were well bundled. Keg Man was wearing only a ripped T-shirt and blue jeans with big holes showing bloody knees, but he didn't appear to notice the cold.

I tried to be amiable. "Whatcha doin' sailor?"

Keg Man looked up startled. He hadn't seen us and he certainly hadn't heard our talking above the din. He was polite and friendly. "Oh, hi guys, how ya doin'? I gotta get this keg back to the store."

"I'm sorry amigo, but the liquor store is closed at two-thirty in the a.m.," said my partner. Keg Man didn't seem to hear.

"C'mon guys help me roll it. I'll lose my deposit."

He wasn't really hurting anybody. It seemed a shame to send the young man to the brig. His troubles over this simple issue would last his whole career, however long that may be. More compelling—my partner and I would have to do paperwork for arresting him. In my mind that settled the matter.

"Friend, we can't help you. You'd better get this keg and yourself the hell out of here."

"C'mon guys, the four of us can do it easy. C'mon!" My partner and I looked at each other and shared a silent laugh. This fiasco had just gone to a whole new level. Keg Man didn't know how many of us there were. Or maybe he did and the word "three" just deserted him, coming out instead as "four."

"Listen friend, we're Shore Patrol. We've got to arrest you for being drunk and stupid if you don't beat it out of here."

Keg Man was incredulous. "You...you guys aren't going to help me? You...you're not very good friends." Keg Man was disappointed in us. We were letting him down.

After some gentle persuading that involved repeating the same argument many times, we finally got Keg Man turned around. Convincing him to stash the noisy empty keg behind some bushes next to the BEQ took more early-morning diplomacy.

He was afraid the keg would be stolen. We convinced him it would not be stolen. We knew it absolutely would be stolen; the deposit was worth thirty-five dollars to any man smooth enough to talk his way through the lack of a receipt. You couldn't swing a cat without hitting a smooth-talking bullshit artist at Great Lakes Naval Training Center. We simply could not allow him to roll the damn thing back up the street, hence our righteous justification for the lie. Even if it did cost him thirty-five bucks.

This was politics at its most local, The Man on the street to the man on the street.

Finally, we had Keg Man pointed in the right direction. As he marshalled his resources for the first step toward his distant bed, my peripheral vision registered a white flash. The Master-At-Arms van had turned the corner and was barreling down on us. It would be trouble for my partner and me if they caught us letting this poor bastard go.

Just as Keg Man was making his first smart move in many hours, I grabbed him by the arm. "Sorry, buddy. Too late. You have to come with us." His troubles just beginning, Keg Man was taken away in the white van. I never saw him again.

In a bureaucracy, and later I would discover in submarines, nearly every inconvenience can be attributed to somebody covering his ass.

10

Nuke School

General requirements for entry in any service school are: Except in the case of a candidate for the Officers Cook's and Officers Steward's School, he must not have a mark lower than 50 on the Bureau of Navigation Standard Test in Arithmetic. Candidates for Officers Cook's and Officers Steward's School must have a mark not lower than 25 on this test.

— *Bluejackets' Manual, 1940 edition*

AT GREAT LAKES I met a girl; we fell in love. Finishing Electronics Technician "A" and Radar school, I was readying to move to Florida for more schooling. Robin wanted to come with me and there was only one practical way to do that. We married in a small church ceremony attended by some of our transient student friends.

After wintering in the deep freeze of Great Lakes, Orlando was a dream come to life. In the weeks before my Nuclear Power

School class started, we reveled in the sun and explored our new surroundings. Disneyworld was nearby, Epcot under construction. We drove to Daytona Beach and parked on the sand like everyone else. We soaked up sunshine and Margaritas. We pale Chicagoans became Florida tan.

We watched a sunrise from a beach on the Atlantic coast and then drove west to see that same sun set over the Gulf coast.

From our front door in Orlando, we watched over the top of the apartment building across the street as Columbia, the first Space Shuttle launched from Cape Canaveral, rose into the blue sky on a Sunday morning. Like an inverted highway flare, the white flame of its engines could still be seen unaided as the TV news announcer told us the rocket was approaching the West Coast of Africa.

Then the fun ended. School started.

United States Naval Nuclear Power School is one of the most challenging engineering schools in the world. The cumulative washout rate for would-be reactor operators during my time in the program, from boot camp to graduating Nuke school, was approximately eighty percent. We who had managed to remain in the program up to now faced a high hurdle in Nuclear Power School. Eighty-hour work weeks became routine.

The first six weeks saw me in mathematics refresher course. Safely completing that, Naval Nuclear Power School class 8203 began.

The class designation stood for the third of four Nuclear Power School classes held in 1982. That class number would stay with me through Nuclear Power Training Unit, known as Prototype. That final training, of course, depended on my graduating Nuclear Power School. The Upper Twenty Percent requirement was now moot. All I had to do to remain in the pipeline was graduate Nuclear Power School.

Risking monotony for the sake of accuracy in the education of a Navy Nuke, please indulge my listing the main course topics at Nuclear Power School. Mathematics, nuclear physics, electrical power theory and generating equipment, nuclear reactor technology, thermodynamics, chemistry, materials science and metallurgy, health physics, and reactor principles. I promise not to further belabor these topics.

Much of the subject matter we studied could be learned at nearly any university. However, since the examples, figures, and technical data used in the classroom were figures actual to classified Navy vessels, the curriculum was classified. All schoolwork and study was done on the premises in the Nuclear Power School building. Security watches were stationed at the doors. Textbooks and notebooks were locked up when leaving for the night. On entry and exit briefcases and backpacks were inspected for contraband.

In addition to the security benefits, this "all work done here" policy constituted another smart training move on the Navy's part. As learning tools, homework and after-class study assignments at a civilian university had proved marginally effective on me. Cramming for tomorrow's test with the companionship of Budweiser and the Rolling Stones was a thing of the past. At Nuke School all study was monitored. Chatter, foolishness, and otherwise unmilitary conduct were not permitted, even after hours, not even a little.

Having a captive student body, the military had the luxury of a different approach to education. In the civilian university I attended, failing to complete assignments was discouraged, but was not necessarily a career ending choice. At Nuclear Power School, failing to complete an assignment was punishable as dereliction of duty. The key word is punishable. The sad perpetrator risked being dis-enrolled from this advanced field and sent away to be useful in the fleet. In place of a scarlet letter he would likely lose a stripe.

The differences between civilian and military education are myriad. My general observations, shared as a layman, are that civilian higher education is in the business of providing educational opportunity for those willing to glean it from the resources made available by the school. Each student will take away a unique set of knowledge along a continuum ranging from zero to one hundred percent, based on which college he attends, his abilities, his effort, and the daily choices of his teachers. The student's grade point average in a particular course of study, while still the universal measurement of academic achievement, is also variable and marginally indicative of his knowledge.

Therefore, a GPA of 3.8 from one university may register a vastly different sense of gravitas than a GPA of 3.8 from another institution.

In contrast, the Navy is in the business of educating everyone in a particular course of study to a precise standard. Everyone will have that standard of knowledge or they will be shunted off to a less demanding career path. There were no do-overs of Navy schools.

The qualification standard was thick and detailed. "Upon successful completion of this course, the student will be able to..." The list that followed was specific and long. If you could do all of those things to the required level of knowledge, you passed. If not, you failed. Simple.

A student with one set of instructors took away the same set of knowledge as students with a different group of instructors. The instructors taught exactly the same thing. They wrote the same notes on the chalkboard. Exactly the same.

Choosing professors known for easy grading was not an option. You didn't choose. The system chose for you. The teachers were all Navy personnel. All exams were graded to a fixed standard. Variable individual opinions of the instructors did not figure in. The instructors taught to a precise standard. There were no individual opinions or teaching styles. This may sound like robotics, but in reality, it was a pure form of results-based education. The results were standard and the standard was high.

Every day of the first month of school, my classroom held another empty seat.

"'Lopes,' where's Dog Chow?" I asked between Thermodynamics and Reactor Principles.

"He's two-point-four-nine," Lopez said. Oh.

We lived or died on our grade point average. Our mantra was "two-point-five, stay alive." Ralston's GPA had dropped below the 2.5 minimum; Ralston was gone.

As Nuke School progressed the washout rate slowed. After three months the class was settled in. Only a few more would leave without graduating.

A normal Nuclear Power School day started at 0700 and ended around 2000. (That's 7 a.m. to 8 p.m. if you aren't hip to the 24-hour clock.) I lived a half-hour away and was rarely home during daylight. My Florida tan faded. The poor time management and lack of self-discipline that stymied me in my short college adventure were a non-issue at Nuke School. My time was managed for me.

The only self-disciplines required were getting to school on time and the will to learn. The learning should have come easy because the subjects were fascinating. It was the pace and volume that was most challenging. My learning rate had to shift into a gear I didn't know it had. The Navy determined to give me a four-year engineering education in six months time. It was like eating my morning Cheerios with a shovel.

As for getting to school on time—you already know that I'm a rule follower.

I spoke weekly to my career counselor. I should say he spoke to me.

"Your last Thermodynamics test score dropped your average a little. Better get that back up or I'll have to put you on M-20," he said one week.

If he thought my grades were beginning to fall off, he would prescribe extra study hours. Most of us were on mandatory twenty extra hours of study per week. Many of us were on Mandatory-40. That was forty hours per week on top of five full days of classroom lectures and homework assignments.

Acknowledging the long hours, it was customary to stand in the back of the room when we became drowsy and keeping our eyes open became a battle. During a class right after lunch, a tough time period for any student, Petty Officer Brad Cherry stood in the back, working hard to stay with the subject. Our attention, focused on the front of the room, was annihilated by the crash of Brad Cherry falling into the metal lockers lining the back wall. The man had fallen asleep standing up, with his notebook in one hand, a pencil in the other. Who knows how long he slept before he tipped over.

We didn't realize that this too was training. Concentrating on procedures and solutions and operating a submarine without sleep for 30 or 40 hours at a stretch was in my future. Even had I known that at the time, there was no turning back.

The final comprehensive exam for Nuclear Power School was a make or break deal. The number and length of equations required to successfully pass the exam made cheating impractical, even if one was inclined to try under the horrific consequences of discovery and capture. The Final Comp had no time limit. Bathroom breaks were monitored; one test taker, one proctor.

As we completed the big test we left the building. An outdoor break area teemed with worried students.

"I hosed up the equation for calculating the shielding thickness needed to attenuate neutrons and gammas," one man moaned. "I think I mixed them up." It turned out that he was wrong, he didn't screw up.

"I got that one," another said, "but I had a brain fart and couldn't come up with the equations for the steam turbine questions. Converting BTU's to pounds-mass per hour kicked my ass." It turned out that he was right; he did get his ass kicked by unit conversion factors.

"Where's Smiley?" someone asked.

"He's still at it."

"Oh man, is that good or bad?" The man could either be working carefully and thoroughly, or he could be in deep shit. Smiley kept at the Final Comp for almost eleven hours. He passed. The student finishing Number One in class 8203 carried a 3.97 grade point average. The Number Two man in our class finished way back with 3.83.

Sixty students began Nuclear Power School as class 8203. I have a photograph of the graduating class. In it are thirty-eight faces, giddy with relief. I'm in the back row, fourth from left.

11

Prototype Training

A submarine sailor is 80% engineer and 20% military man. In the case of a Nuke it's more like 98% to 2%, and that irritates hardcore military men.

— *The Submarine Sailors' Imaginary Book of All Knowledge*

NUCLEAR POWER SCHOOL checked off as a success, Robin and I moved on to Prototype training in Windsor Locks, Connecticut. A few NPS friends made the move with us, others went other prototype facilities.

"Jones, which Prototype are you going to?"

"S1W in Idaho Falls, baby!"

"You sound happy, are you from there or something?"

"Nope, I'm from Northern California, but I figure Idaho Falls is closest to the West Coast and I'll have a better chance of getting assigned to a West Coast boat after Prototype." In a perfect world there might have been validity in that line of reasoning, but this was the Navy. What seemed reasonable to we students had little basis in the larger scheme.

The S1C prototype was the back half of a submarine mounted on land. Airlocks at each end of the cylindrical hull provided yet another level of radioactivity containment should the unlikely event of the unthinkable occur. Instead of a propeller, a braking device was installed which simulated the resistance seawater and inertia would place on a propulsion plant at sea. Originally a test platform, the S1C plant was now a training facility. But before she was retired in 1993 after training some 14,000 Navy men and women, S1C would contribute engineering knowledge to the always relevant subject of extending reactor vessel life.

Prototype's *raison d'être* was to provide an effective segue from theoretical knowledge to practical application.

First on the training agenda came a six week classroom phase. We learned design, theory of operation, and maintenance procedures of in-hull systems. We laid hands and eyes on spare parts and broken parts and learned how the most expensive equipment in the world could fail.

After classroom phase came in-hull phase. Students worked 12-hour shifts. After seven days of midnight shift (Mids), we got one day off. Then seven days of the afternoon shift, (Swings), and two days off. Seven days of Days followed and then—glory, glory—four consecutive days off. This shift cycle repeated for my six months as a trainee and the following twenty four months as staff at prototype.

"How's it going, Ken?" I asked a friend.

"Living for that four-day," he said. You know it brother.

Rotating shift was toughest on Mids. You can't bank sleep. Four o'clock in the morning was four o'clock in the morning. No one slept well until day four of the week-long shift. Even then it was nigh impossible to get more than six hours of restful sleep. Dark towels over the windows helped a little, but the tiniest crack of light seeping through was as distracting as a 747 landing in the driveway.

Just as my body became accustomed to the sleeping hours, the rotation would roll along and move my crew to swing shift.

Climbing into the car one day I said to my carpool mates, "Let's go to school."

"Bullshit on 'school,' man," said the driver. "We are Navy men, petty officers second class, pay grade E-5. We take home a paycheck. We are in an advanced engineering program with a schedule so demanding that no labor union would agree to anything close to it. We don't go to 'school' anymore."

"Alright then," I said. "Let's go to work."

Back at Nuclear Power School, we had been a single homogeneous unit of students studying all aspects of nuclear powered propulsion and electrical generation. At prototype, we separated into our individual ratings. Electricians' mates became Electrical Operators, machinist mates became Mechanical Operators, and electronics technicians became Reactor Operators. A few machinist mates became Engineering Laboratory Technicians.

We all continued to learn the plant as a whole. Cross-training was a staple of our Navy Nuke diet. We had to prove our knowledge of, and ability to do, the other guy's job as well as our own.

Electrical Operators, EOs, were responsible for the electrical generation and distribution end of the enterprise. Sharing an electrical grid between multiple generating machines is science of course, but in the quickly changing electrical environment of a submarine, it is also an art.

Reactor Operators, ROs, operated and maintained the reactor plant and its associated control systems. Under our list of responsibilities were the nuclear instruments, primary plant instruments, protection and alarm systems, rod control mechanisms, et cetera, et cetera.

Mechanical Operators, MOs, owned about everything else in the engine room. The steam plant, the propulsion plant, and all the associated systems were operated and maintained by MOs. Hydraulic plants, turbines, water making, seawater systems, the emergency diesel—if it was a hot, dirty job, involving steam, water, or oil, it belonged to the MOs.

The ELTs were part of Mechanical Division, but instead of working the typical M-DIV gear, they had responsibility for water chemistry in the primary and secondary plants. ELTs also handled the radiological issues. That included safety and radiation exposure monitoring of the humans inhabiting the boat. ELTs specialized in decontamination procedures. Particles invisible to the human eye were the ELTs' bailiwick.

Studying a technical manual on the Sharples lube oil purifier, a classmate looked up. "Wouldn't it be great if I could take this home? Stretch out on the couch with my feet up and a cold beer in my hand, and study this like a normal human?"

Nice try pal, but no.

As in Nuke school, all textbooks and technical manuals were classified. This would apply to every book and manual on real submarines as well. Nothing left the site.

Work shifts included eight hours in-hull. We stood watch "under instruction." We learned to operate the plant, do maintenance, and traced out in-hull systems. Every pipe, every valve, every breaker, switch, and fuse had to be identified and discussed. After our crew's eight-hour watch-standing shift came four hours of study in a common training area with rows of study carrels. The study area had a library with all the books and manuals we needed to learn each system and each component intricately.

One day my Training Advisor hammered on me for lack of progress toward qualification.

"I thought this program was self-paced," I pointed out.

"You believed the sales brochure, you dumb fuck," he said. "Marketing departments never know how anything really works."

The term "self-paced" was somewhat misleading. There was a scheduled date by which all students must be qualified on every system and on the plant as a whole. The program was structured and monitored, but what I chose to study today was up to me. There was a minimum acceptable pace and we were welcomed to progress faster. That's self-paced the Navy way.

This process was a warm up for the real world of qualifying on a nuclear propulsion plant on a ship or submarine in the fleet. There we would learn under our own horsepower. At prototype we learned how to learn a submarine.

Motivated quick-learners leaned into the task. These "hot-runners" had opposites; "slugs."

When we thought we knew a system thoroughly, we asked a staff instructor for a checkout. He would test our knowledge. Usually this involved drawing on the chalkboard a diagram of the system, showing all the components. Then we listed the specifications of each component in the system. We were quizzed on every aspect of every component, as well as the system as a whole.

"What type of valve is that?" a staff instructor would ask. "Why that type?" "What's it made of?" "Why that material?"

Flow rates, operating temperatures, pressures, what type of hydraulic fluid was it and of course, inevitably, why? What were all the ways of operating the system? What if electrical power failed, what then? On and on it went until the instructor was satisfied that we knew what we were talking about. He wasn't working off the cuff; he had the qualification standard in hand. This Qualification Standard was the two-inch-thick book each student carried that said, "The student will know…"

If we knew the system to the satisfaction of the Qualification Standard, the instructor signed our qualification card in the block for that system.

If we did not know our stuff, the instructor would, in good Navy fashion, berate us for being lightweight know-nothings and wasting his time.

"Come back when you know something, Sluggo," was a common refrain heard across the training area. At one time or another it was said to all of us. Some heard it many, many, many times. The workload of learning was heavy enough that even the brightest occasionally tried to sell a little blue sky and get a signature without adequate knowledge. We students were sailors too, by Rickover. The instructors—experienced sailors—would have been disappointed by our low commitment had we not pushed the envelope.

An instructor might also assign look-ups, as in: you almost have it; you are weak here and here. Look up the answers to these questions, and then I will sign your card.

Each signature on our qual cards carried a point value. Progress was monitored, and could be self-monitored, by achieving a prescribed point count each week. Back in Nuclear Power School our grade point average was everything. At Prototype our lives hinged on how many points we acquired each week.

As with Nuclear Power School, the pace of expected learning was challenging. At some point everybody was in trouble with their point count.

We trainees stood watches in-hull, just as we would in a ship or submarine at sea. An instructor, standing watch alongside the "Under Instruction" watchstander, confirmed every move before it was made.

"If this meter reading rises to that mark, what does it mean and what are you going to do?" the staff instructor might ask.

"It means that it's time to shift the Low Pressure Cut Out interlock switch to Cut Out. I'm going turn this switch to that position," I might say.

"In which direction are you going to turn that three position switch?" the staff instructor might ask.

"To the right," I might say.

"Which way is 'to the right'?" the staff instructor might say.

I mimicked turning the switch.

"Correct," the staff instructor might say. "If you turn it any other way I will rip your arm off and beat you over the head with the bloody stump."

It was astounding how many students would then proceed to turn the switch the wrong way and accidentally shutdown the reactor by inserting a safety interlock that was no longer applicable. This was not catastrophic to the machinery since the reactor plant is designed to safely shutdown on occasions such as this, but on a submarine at sea it would be like throwing the gearshift in neutral while trying to get somewhere. Under certain circumstances that could be catastrophic.

During slack times in the watch, the staff instructor ran a continuous battery of "what if" scenarios at the student watchstander. UI watches were rigorous learning experiences.

We learned maintenance procedures in the same fashion. Students performed the procedures monitored by staff instructors.

An emergency such as fire, flooding, electrical malfunctions, machinery abnormal behaviors, are known as casualties. Casualty drills were run by a team trained in simulating likely scenarios. The drill team opened breakers or pulled fuses to produce false indications. Black plastic trash bags simulated smoke. These simulations often smacked of grammar school theater but they got the job done. The staff instructors monitored student watchstander actions and prevented actual damage to equipment or persons. In this way we practiced what to do when things went awry. We learned to anticipate trouble and we learned what to do about it when it happened.

The skills and abilities of the staff were also tested and evaluated regularly. Our instructors were instructed. Staff drill sets were organized by the Staff Training Group. If an instructor failed to excel, he would lose his instructor credential until he upgraded himself and proved his knowledge.

Prototype was where it all clicked for me. All the example problems, all the equations and theories of fluid flow, thermodynamics, fission and chain-reactions, particles and energy, radioactive decay, turbines, pumps, motors, generators, chemistry, physics, Newton, Bohr, Rutherford, all the mathematical *stürm und drang*, finally made sense. I could see the interwoven relationships of all the systems.

The "submarine" needed more speed; a higher bell was rung up on the Engine Order Telegraph. The Throttleman opened the throttles to the propulsion turbines. More steam poured onto the turbine blades; the turbines increased their revolutions per minute. Since more steam was being sucked out of the steam generators their pressures and temperatures dropped. This increased the rate of heat transfer from the primary coolant system to the secondary steam system through the steam generator. The increase in heat transfer out of the primary caused the primary loop cold leg exiting the steam generator to lower. Colder water entered the reactor vessel and the fuel core. The colder water was denser thereby increasing the moderation of fast neutrons from fission into slower thermal neutrons more likely to be absorbed by a U-235 nucleus. More neutron absorption increased fission rate. Increased fission raised the output temperature of the primary coolant, the "hot leg." Hotter water entered the primary side of the steam generators putting more BTUs into the steam generator primary side to be transferred into the secondary side to match the increased steam demand. The primary and secondary systems settled into a new steady-state level of equilibrium.

The hum of the pumps, the whine of the turbines, and the muted roar of steam through insulated piping made sense to me. It was mathematical, it was beautiful, a wondrous ballet of physics; first this, then the other, resulting in that—every time.

Readily grasping the physical concepts of the nuclear propulsion plant, I quickly qualified as a Navy Reactor Operator. Finally, after two and a half years of schooling, I was a bona fide Navy Nuke.

12

Getting There

'Cause only one thing counts in this world: get them to sign on the line which is dotted.

— *Glengarry Glen Ross*

GRADUATING PROTOTYPE TRAINING high in the class standings, I was invited to stay on as an instructor. As with every good deal in the Navy there was a catch: accepting the job required extending my enlistment.

Navy Nuke wanna-be's enlist for six years. The schools I attended on the way to becoming qualified as a nuclear operator took two and a half years. The Navy wanted a minimum of 36 months of sea duty in exchange for all this education. We were told the Navy spent one hundred thousand dollars training each nuclear operator and the Navy expected a return on its investment. An additional 24 months of obligated service was required to stay on as staff, bringing my total to eight years.

In business this is known as the up-sell. Your hand is already in the man's pocket, try to grab a little more while you can. He's already said yes to buying the car; now sell him the custom wheels, the upgraded stereo, undercoating, a spare tire, and an extended warranty.

Wanting to start a family, Robin and I decided that staying on as staff was the right play for us. Accepting the position and added time obligation, I would be landlocked for at least two years. That would be good for us. It would also be good for baby Hannah who didn't yet know that she was heading toward her birth day. Not the anniversary, the real thing.

Robin and I lived in Springfield, Massachusetts, about 30 miles north of the Windsor site. Whichever route you took, the drive to work usually took forty-five minutes at 5:30 in the morning. It could take as much as an hour and a half on Friday afternoons. That was the day shift dilemma. The day shift work hours were more conducive to pretending to have the nine-to-five life enjoyed by our civilian friends, but the commute sucked. During backshifts the trip wasn't so arduous.

Car pools formed and evolved as people moved into various operating crews, each with a different schedule. Now a staff instructor, my work shifts were only 9 hours long. The first hour was spent in a briefing, usually done by one of the General Electric test engineers.

I have never been what one would call a "morning" guy. I didn't mind being up early, once up and rolling. Witnessing the dawn always seemed a gentle reminder that each day is a fresh start. I'm just not particularly talkative before nine. My carpool mate Gary shared similar traits. We enjoyed our coffee and the lightening of the eastern sky on the drive into work, never saying more than a few words.

"Hey," Gary would say as I climbed into the front passenger seat of his car at 5:15 in the morning. Gary had been to sea on a submarine and was now doing shore duty at S1C. Senior to me in every respect, he also had a nice car.

"Hey, yourself," I'd reply.

"Everything good?"

"Yeah. You?"

"Good." We would drive for the next 44-1/2 minutes with the radio volume at a mutually acceptable early morning level. No more words were spoken. The car parked, we wordlessly walked to the guard shack. After passing through metal detectors, radiation detectors, and running our lunch boxes through the x-ray conveyor, we entered the Windsor site.

"Have a good day."

"See you."

That was it. The perfect morning commute. A tall cup of coffee, smooth jazz on the radio. A beautiful sunrise. Minimal conversation. Just right.

In time I moved to a different shift and my pleasant carpool with Gary dissolved. One icy winter morning, my new car pool mate, Randy, drove around a bend on Interstate 95 to see a line of red brake lights in front of us.

Standing on the brake, he one-handed the non-power-assisted steering 1968 Dodge ¾ ton pickup truck, horsing the oversized steering wheel around in one direction, then winding it around the other way as the rear end of the pickup fishtailed. We came to a stop six inches from the rear bumper of the car in front of us. The pickup was lined up straight down the road as if nothing out of the ordinary had happened.

"Nice," I said.

Randy nodded and made a sharp noise that sounded like air let out of a balloon. "That was pretty close," he said. End of conversation.

Hours later, after morning had worn off, I bragged on Randy's skill in strong-arming the old Dodge all over the road, saving the most spectacular part of the story till last.

"He left-handed that big ol' wheel three turns this way, then he open-palmed it three turns the other way. If the car in front of us had another coat of paint on his back bumper we'd have wiped it off for him."

Then the crescendo. "AND during the whole thing, the smooth bastard never spilled a drop out of the coffee cup in his other hand!" The crowd around the coffee pot nodded and chuckled in appreciation of the skill and respect for good coffee. Some things are more important than others.

Randy told me later that his ribs were sore for a week from the effort.

We took on a new carpool member. Jim was one of those people who always woke up bright-eyed and bushy-tailed. Cheerful and chirpy at five in the morning, it was all we could do to tolerate that much joy at that hour. He was a fine guy; it was just too early for Randy and I to be cheerful. We stared out the windows as Jim chattered.

The driver chose the music, a self-inflicted rule we now regretted. For one entire week, Jim played the soundtrack to *Oklahoma*. It was his new favorite and he played it from its beginning every day. Randy and I stared out the window for 45 minutes.

On the last day of Jim's turn to drive, "Surrey with the Fringe on Top" was in full swing, with Jim on lead vocals. We had heard this performance twice a day for the last seven days.

With a tone of high curiosity Randy said, "Could I see that tape for a second, Jim?" Jim plucked the cassette from the player in the dash and handed it over. Thrilled at the possibility of having converted Randy into a fellow Rogers and Hammerstein aficionado, Jim enthusiastically spouted trivia about the musical.

Without glancing at the cassette, Randy rolled down the window and flicked the tape out into the darkness. He rolled up the window and looked straight ahead. The car was wonderfully quiet for the rest of our ride.

Under normal circumstances, an enlisted nuke assigned as staff at a nuclear prototype meant instructing. A new class of students arrived every ninety days. However, S1C was preparing to shut down for a refueling overhaul. Mine had been the last class of students for the next two years. Instead of instructing, the staff would overhaul the prototype. Real work, Navy Nuke work. The business of training would start again after the refueling overhaul was complete.

Refueling the reactor was a massive effort. Replacing the core containing the nuclear fuel essentially required disassembling the whole thing. A hull cut was made and lifting equipment was rigged down through the hole.

Most of the heavy lifting, both figuratively and literally, of the overhaul and refueling was done by civilian contractors. Heavy machinery moved in and temporary offices and workshops were set up for the horde of new workers. They had no trouble finding places to live for the next two years on their generous per diem allowances. We despised the civilian workers new customized pickup trucks while most of us Navy guys drove second or third-hand cars.

It was a busy time that was exhilarating. Every day brought a new adventure; a new test to perform, a new piece of equipment to figure out. We became experts by poring over technical manuals and teaching each other what we'd learned. Great discussions and arguments occurred over specifications and principles of operation.

"The new Protection and Alarm circuitry works like so…" someone would say. Another someone would counter.

"No, it doesn't. The new P&A works like this." Tracing out schematic diagrams of the electronics, the facts were sorted out.

The setting of Design School was that of a college classroom. The "professor" was an engineer from the company that made the equipment that was the topic of the day. It was fascinating to hear a Navy electricians mate first class argue face to face with the Westinghouse PhD who designed the primary shield tank. The PhD may have been correct theoretically about the particular issue, but the petty officer had operated the equipment as a part of a complex assemblage of machinery. Both men were right, but in different contexts.

As S1C staff, even with the workload of a complete refueling overhaul of the plant going on, life was good. Even though we were a U.S. Navy installation, the real Navy seemed worlds away. We worked with more civilian technicians and engineers than we did with Navy people. We lived in civilian apartment complexes and shopped in civilian grocery stores. We felt like civilians, just guys with regular jobs, and that felt fine.

Soon enough we would be assigned to a sea-going submarine and all sense of normal would disappear.

13

Chit Storm

All log entries are made with a ballpoint pen using black or blue-black ink.

— The Bluejackets' Manual, Twentieth edition

PERSONAL REQUESTS IN the Navy were conveyed up the chain of command on a five-by-seven-inch document, NAVPERS 1336/3, known as a Special Request Chit. Holding a stack of blank request chits felt like holding a stack of money.

A service member may, for instance, request a day off to attend his sister's funeral. In the civilian world, this day off might be a given, assuming that one is not an astronaut in orbit. Not so in the Navy. Nothing happened in the Navy without a valid Navy reason and the proper paperwork to support it.

Your request chit passed up your chain of command, each supervisor in turn marking his recommendation. The top of your chain of command decided the final "approved" or "disapproved." If your request chit came back approved, you had the day off.

If your requested was disapproved, most often it would be an issue of available manpower requirements. It might also be disapproved because you had asked for too many things too often.

There wasn't supposed to be, officially, but in reality there was an element of punitive payback in the disapproval of some requests. An enlisted man requested time off to consult with a Navy attorney over an impropriety he had been charged with by the Navy. The man's request was denied on the grounds that he had been bad and was therefore undeserving of a day off. After pointing out that the Navy was denying him counsel in its own proceedings, clearer heads prevailed. He got his day.

The special request chit was also used for lighter subjects. In 1985 Congress passed the Graham-Rudman-Hollings Act in yet another attempt to bring government spending under control. The Navy decided to help. A program was instituted whereby a cost saving suggestion made by a service member was rewarded with cash.

A Navy man suggested that in an infrequently used storage building at S1C, the lights could be turned off when no one was in the building. His genius was rewarded in the form of a check equal to ten percent of the annual savings to the defense budget that his suggestion had caused. This was a game that only recognized first place finishers. The guy who suggested that the same storage building was actually unnecessary and could remain unheated as well, was too late. Certainly this idea would have occurred eventually in the natural course of organizational stewardship. The S1C plant had been in operation since 1959. It was now 1985. Some good ideas need time to germinate.

This business of saving money for Uncle Sam was heady stuff. For several months a flurry of ideas were tossed around as sailors brought all their creativity to bear in the interest of improving their personal cash situation. Some ideas did not go far up the chain of command before being shot down. Many of the more entertaining denied requests found their way onto a common area bulletin board.

A recent article in Time magazine described the massive annual cost to the American taxpayer of housing death row inmates. Many of these inmates had been on death row for decades.

Machinist Mate First Class Chris Morelli did the math. He put in a request chit promising to save the American public millions of dollars. He had calculated the cost of commercial airfare to send him to every major U.S. prison—first class, of course—and the cost of hotels and a generous per diem chow allowance. He added in the cost of one .45 caliber pistol cartridge per death row inmate. As a gesture of fairness he was only including prisoners who had already been denied on appeal.

For the convenience of the government, Morelli had already calculated the savings and the amount of the reward check to be made out to him. In courtesy to the clerk who would make out the payee line, he printed his full name clearly in block letters. An asterisk pointed to the annotation that he would provide his own firearm for this mission.

Request denied.

In August 1984, President Reagan announced NASA's Teacher in Space Project. The program was a grand public relations effort to honor teachers and inspire students. A selected teacher would travel into space on an upcoming space shuttle mission. Applications of deserving teachers were requested.

Electrician's Mate Second Class Roger Jarvis thought the field was too limited. His chit read: "Respectfully request to be the first Navy Nuke in Space." His immediate supervisor, a terminally exasperated man named Walters, had recommended approval of Jarvis's request and added the underlined words "please expedite approval" after his signature.

Request denied.

Political correctness was a consuming doctrine of the times. Every organization in the country, large and small, was careful to keep the wording of everything as neutral and inoffensive as possible. Personal sensitivities were high on every subject. We, the American public, were accustomed to social and economic groups expressing their jilted frustration, but now everybody was shoehorning in on the act. Everybody became their own minority and their own favorite cause. Everyday people were learning to make a business of indignation.

A civilian friend confided to me that men in his office refused to speak to the females in the company except as required in the strictest business applications. This silent treatment had precipitated out of an incident where a man had complimented a coworker on her attire and found himself in danger of losing his job.

Certainly there is nothing to be gained from a poor choice of words and much put at risk. However, the farther one is from a seat of authority, the less conscientious one tends to be in recognizing the full potential for harm of a seemingly insignificant action. The first sieve is the coarsest.

The staff instructors at S1C were mostly pay grade E6 or above. Students were gone, instructors were overhauling the plant. They were in that comfortable position of having enough seniority to simply do their job and mostly be left alone otherwise. The training unit, now an industrial overhaul enterprise, was bare minimum on military pomp and circumstance.

I had not yet been to sea, but the irreverent attitude of the salty dogs who had been there rubbed off on me quickly. I admired them. Even though they might hate the bureaucratic structures prevalent in the Navy and more so in nuclear power, they were completely dedicated to doing their jobs properly and insisted on professional competence from those around them. It was a matter of personal pride. They despised any obvious display of rank or authority. There was no need for that bullshit. Everyone knew who was in charge of what and who. Stomping around shouting and chest thumping was demeaning to everyone and marked a person as lacking problem solving abilities.

During my thirty months at S1C there were three nuclear-trained women instructing there. Previously women had been admitted into the nuclear power program, but that had ended. Women Nukes still on active duty were grandfathered into the program. "Grandmothered" seems more accurate, but might be demeaning. Political correctness can be a bitch. A bastard. Whatever.

However, women were not allowed by Congress to be assigned to combatant vessels. It so happened that all nuclear powered ships and submarines were combatant vessels. So the ladies took the shore duty and the men went to sea.

We men respected and thought of the women Nukes with the same professional respect or disdain as we did other men. Everyone who was competent and carried their share of the load was respected. Anyone who was considered to be marginally competent—as in having to be watched when doing something important or tricky—or shirked their duties, was not regarded highly. Actually, they were despised.

We men hated that the women filled all the nuclear shore duty and we had to pick up female slack in the sea-going department. That was not their fault, of course, and they may have hated it as well. We also hated that when a woman was nearby we had to be cautious with our language and conversations. We weren't bent on disgusting behavior, but we were men working together. We knew each other and had a vocabulary that worked for us. And—we were sailors.

"Hey numb nuts, get your ass over here and bring Petty Officer Shit for Brains with you. I've got a really shit job for you two. You're gonna love it! " was perfectly understood and prompted immediate action. No offense was meant or taken. The blow of the distasteful job was lessened by the levity.

But with a woman around, we were required to speak a foreign dialect. "Petty Officer Jones, come here and bring Petty Officer Smith with you. I have a task for you that you will not enjoy," just didn't get the same response. It was staged and colorless. We knew what he really wanted to say.

The S1C plant at NPTU Windsor was manned twenty-four hours a day. Four crews operated in a rotating shift schedule. Every day of the year, three crews worked eight-hour shifts, day shift, swing shift, the midnight shift. One crew out of the four was "on the beach."

Each crew had enough representatives of each division, E, M, RC, and ELT, to operate the plant and do the required maintenance. Each division had their own workspace in a building off-hull that was shared with their division on other crews. This workspace—the cage—was one part repair shop, one part tool storage, one part technical library, and ten parts lounge. Old office chairs had been scrounged and there was the comfy air of a clubhouse even though all the furnishings were painted gray.

Double-door metal bookshelf units loaded with equipment manuals and reactor plant manuals lined the walls. On the inside of one of these normally-shut bookshelf doors hung a calender.

It was a *Playboy* calendar. The newest version, brought in each December and ceremoniously installed in the book locker by Bob Longworth. We applauded and toasted the newest pinups with bad Navy coffee. The Navy is all about traditions. RC division was all about carrying on this particular tradition.

Our calendar posting was perfectly innocent with no malicious or perverse intentions beyond our own pleasure of basking in the presence of beautiful women scantily clad. It was tasteful. It was art.

The young beauty posing for the December calendar wearing a Santa hat and nothing else, wanted to be a civil rights lawyer. She wanted to save the whales. She loved children and puppies. It said so in her biography. She had a beautiful signature, flowing and smooth. This was a nice girl.

This was what we were ready to go to war for, damn it all. Baseball, apple pie, and the American way. The NFL, right and wrong, the designated hitter, and this young lady's freedom to express herself. Anchors aweigh shipmates! We will crush whatever forces of evil conspire to curb this young woman's dreams of a bright tomorrow. Long live the First Amendment.

One Monday morning I needed a calender for some paperwork. Opening the book cabinet door I found a half-circle of torn calendar underneath the piece of Electric Boat green tape that had previously attached the calendar to the cabinet. Miffed at this inconvenience and loss, I put a yellow sticky note under the calendar shred that read: "What low-life scum took the calendar?"

On arriving to work Tuesday, I checked for a response. A yellow sticky left by someone on another crew was hanging beneath mine, neatly lettered with: "ET1 Bissonette took the calendar."

Who was this Bissonette guy? Obviously he was an FNG who knew not the sacred traditions of Reactor Controls Division, S1C reactor, NTPU Windsor.

I placed a yellow sticky under that one that said: "Tell ET1 Bissonette to get his own fucking calendar and put ours back."

The next day, Wednesday, another note under mine. "HE is a SHE." Oops. I didn't see that coming. Still perturbed at the theft, I placed another note on the lengthening line of yellow squares.

It read: "SHE should get her own *Playboy* calendar and put ours back." It seemed a fair request.

Thursday, the first thing I saw when I opened the cabinet door was the absence of yellow stickys. A sense of foreboding washed over me. This little game should have continued. This could be trouble.

A minute later, Dave, my Leading Petty Officer and my direct boss, handed me a note. It read: "ET1 Heckman. See me." It was signed ETCS Latourelle. My stomach tightened. Yep, trouble. My presence was requested by my boss's boss, the Senior Chief. This was the Navy equivalent of being sent to the principal's office.

Hmmm. Perhaps I had gone a yellow sticky too far.

The Senior Chief sat at a gray metal desk. A green blotter padded the desk top so that a man could write without punching holes in the paper. The rubberized desktop was cratered and holed by previous generations of bored Navy men wielding sharp objects. In the center of this pad, squarely framed in the green blotter, lay a single piece of paper. It was a special request chit.

This was not good. I had not put in any requests recently. I'm sure the color drained out of my face. It usually did whenever the cold stare of unfriendly authority fell upon me.

ET1 Bissonette was accusing me of sexual harassment. I was suddenly aware of a fierce itch under my shirt collar, as if a rope had roughly tightened around my neck.

I felt my face flush, blood rushing up until my head was hot. My ears felt as if they were glowing red. My mouth went dry, I couldn't think. A hundred pounds of words fluttered in the tornado going on inside my head; I couldn't latch onto any single word in order to speak it. When words did come, they came all at once, tripping over themselves.

"Wh-What? You gotta be kidding me! Senior Chief, I've never laid eyes on this woman! I didn't even know she was a woman until yesterday!" The Senior Chief looked confused by this statement. I retract that; Chiefs would want me to say that Navy Chiefs are never confused, by definition. Fine. The Senior Chief looked...quizzical.

I gave a Reader's Digest explanation of the string of events that had brought us here. Between my explanation, him slowing me down to explain my explanation, and her detailed addendum to the request chit, the Senior Chief pieced together the big picture.

Bissonette believed that someone—me by default as the writer of caustic yellow sticky notes—had strategically placed the offensive calendar to torment her personally. She was the only woman in RC division. To her, there was no other explanation.

She was new to the facility and perhaps sought to declare herself right up front as not taking shit from the guys. I could respect that. It's possible that being a woman in a Navy specialty force populated 99.99 percent by men had put an understandable chip on her shoulder. At the very least she was likely a little jumpy. She was, after all, an attractive woman swimming in a pool of male sharks. In a flattering yet tasteful bathing suit I'm sure.

"Is that it? Is that what this is all about—a goddamned calendar?" The Senior Chief's stern eyes sank into a look of dejection. He deflated about ten pounds per square inch. It was now his stomach that turned sideways as he realized the ordeal that lay before him. Instead of passing an egregious issue up the chain for formal legal action, he was going to have to resolve this uncomfortable issue himself.

Everyone saddled with such a chore hates it. This was the case of the stupid smart-ass sailor boy vs. the righteously angry woman. A bet placed on the sailor boy would likely be money lost.

The Senior Chief groaned. "Christ."

He thought for a moment, and then stood, all-business again.

"Petty Officer Heckman, you will steer clear of Petty Officer Bissonette until I get this chickenshit nonsense sorted out."

"Aye, aye, Senior Chief."

"Get out."

My reptilian brain instinctively had me out the door as close as possible to instantly. The less the Senior Chief saw of me the longer my DNA would survive on the planet.

The Senior Chief called after me, "And no more notes! About anything! Ever!"

Walking away down the hall I could still hear him calling out specifics. "Zero! Zip! Zilch! Nada! None! Ixna on the otesna!"

The Senior Chief managed to persuade Ms. Bissonette that I was simply a jackass and meant no offense. She accepted my sincere apology coolly. Icily is more accurate. For the rest of my time at the site I instinctively avoided her as I would a mother grizzly bear with cubs.

She was probably a nice person. I'll never know.

14

Party Hardy

I drink not because I think alcohol makes me more interesting; my drinking makes you more interesting.

— *The Submarine Sailors' Imaginary Book of All Knowledge*

TRADITION HELD THAT we partied on the last night of swing shift; the event cleverly named End O'Swings.

"Going to End O'Swings?"

"Sure, man, I'll be there."

Connecticut liquor stores closed early, so an off-watch person was sent out before the end of the shift to acquire alcoholic provisions in Massachusetts. Trust Navy men to circumvent authority.

After shift we gathered at the designated house or apartment. Sometimes the man volunteered his residence, but this was not a requirement. His house could be volunteered; he could stay for the party if he wanted. We chose locales where noise wouldn't cause a problem. A lasting problem for the tenant was of no concern to us; a problem that might abbreviate the party was the important issue.

We held an End O'Swings at a rural home rented by three Navy electricians. Located near a Connecticut farm with cornfields all around, it was perfect for the party. There were no neighbors near enough to fuss about loud music.

Hours later, the beer keg spitting foam and the sky growing light, the party was breaking up. Those first to leave hustled off when the law came up the driveway.

No one cared about the noise, but apparently they cared about the corn. The sheriff's deputies questioned the slow-to-leave about the crop circles that had appeared in the cornfield during the night. Of course, no one knew anything. The two deputies continued to inquire. They didn't seem bothered by the partying. They persisted in their belief that we had knowledge of the incident and were not coming clean. The law officers were cordial; they seem to be enjoying themselves. They were certainly in no hurry to leave.

"Whose car is that?" The deputy pointed.

The black Chevy Malibu was the pride of one of the electrical operators, Steve. He had passed out on the sofa an hour ago. Someone roused him; he stepped onto the front porch shielding his eyes from the dawn. The morning sunlight seemed very loud.

"Do you know anything about crop circles in the field across the road?" the deputy asked Steve.

"No sir." Good man, always revert to politeness with authority figures. Especially when you may be in trouble.

"Is that your car?"

"Yes sir."

"Did you drive your car last night?"

"Yes sir."

"Did anyone other than you drive that car last night?"

"No sir."

"Son, can you tell me what is hanging off the bumper there?"

From each corner of the front and rear bumpers of the Malibu hung broken corn stalks. From the side mirrors and windshield wipers too. Remnants of corn stocks were plastered into the wheel wells and jammed into the tire tread. Corn stocks were twisted around the axles and undercarriage of the car. The shiny black paint had green streaks and smears on it. If any car looked like one that had made crop circles in a corn field, it was this black Chevy Malibu. Steve stared unbelieving at his car.

A faraway look came over his face as he dredged up a distant, misty memory. Almost…almost got it…Bing! A moment of clarity chopped through the jagged haze left by seventeen beers, four shots of Cuervo Gold, a shot of Jaegermeister, and two pints of Long Island iced tea.

"Oh," Steve said, remembering. "Oh, yeah…that."

The sheriff's deputies' crime solving instincts were spot on. They had their man. It could have gone hard on Steve, but he had the good fortune to screw up at the right time. It was late in the season and the last of the year's corn had been harvested a few days before. The farmer was mad, but not too mad.

Steve ended up paying a fine and doing a hundred hours of community service on the farm he had vandalized. He and the farmer ended up friends.

Later that winter, in the middle of February, four bachelor ELTs who shared a classic old colonial house, threw the party of the century. They covered the walls with bed sheets painted with palm trees. Every light fixture sported a heat lamp. Four inches of sand covered the first floor of the old house.

The theme was "Winter in Jamaica," and Bob Marley sang out on the stereo. We dressed in shorts and aloha shirts. Everyone wore sunglasses, straw hats, and sunscreen on their noses. A whole pig roasted in a fire pit dug into the snow covered lawn of the back yard. Rum drinks with umbrellas filled every hand that didn't have a beer in it. The party continued 30 hours straight. New partiers came as crews rotated off duty. People talked about Winter in Jamaica for years.

When their time at S1C ended, the four house mates were indignant when their security deposit on the house was not returned.

They felt violated and sought legal representation. They met with their lawyer after he had spoken to the landlord.

The four sailors left the lawyer's office quickly and quietly. There had been words about a basement full of sand that had leaked down between the floorboards of the old house, and holes dug through the backyard sod big enough to bury a body.

Fortunately for the heroes of Winter in Jamaica, they all moved to new duty stations out of state. They went minus a hefty cleaning deposit and an extra month's rent.

15

Mare Island

Every man thinks meanly of himself for not having been a soldier, or not having been at sea.

— Samuel Johnson

MY TIME LEFT at S1C growing short, a dark-suited man arrived at the site. In a borrowed office he flashed official looking identification tucked behind a clear plastic window in a black leather wallet, expertly flipping the top shut before I could get a good view. He asked if I was interested in being a reactor operator on a Special Projects submarine. I had no idea what that meant. His explanation was so vague that I don't recall a word and knew nothing more than when I asked the question. He was that good. Today he probably runs the CIA or perhaps even a real super-spook organization: the kind with no acronym, only a code name.

He queried me about run-ins with the law. He wanted to know every traffic violation. He was interested in my family background. He wanted to know every country I had ever visited.

He was especially interested in Russia. He asked about my heritage. Norwegian and English I said. At least that's what has been reported to me. I explained that I had been purchased at four days of age. My parents' genes ran strong toward females; any males added to the clan would have to be traded for or bought outright. Our family doctor had known a guy.

I'm sure I was swapped in the hospital parking lot in the dead of night for a gym bag full of cash. The dark-suited man kept a masterful poker face during this recounting of my history. Or perhaps he was truly that bored with the tale.

He continued his battery of questions. Any Russian in me at all, even a little bit? I mentioned that my grandfather had done some traveling, so I may have a few cousins of which I was unaware. The dark-suited man did not smile. He wrote. After I told him I was joking about the cousins, he wrote again. I stopped joking with the dark-suited man.

I couldn't help but be intrigued by the prospect of Special Projects. I had joined the Navy for adventure and anything with Special Projects in its title clamored for my attention. I said yes, if they would have me, I'd do it.

The job required a minimum of 36 months of active duty service. Assuming I left S1C as scheduled, I had only 30 months left on my contract. Once more I extended my enlistment to take the job.

Accepting the assignment on USS Seawolf, home ported at Mare Island Naval Shipyard in the far-off mystical land of California, was contingent on obtaining a TS-SBI security clearance; Top Secret with a Special Background Investigation. During the months until that was completed, I would stay where I was and do my part for S1C.

Eventually I received orders to *Seawolf*. En route to that new duty station, I was to attend a week long school at Naval Submarine Base New London. LMET was the acronym for Leadership Management Education Training. Up to now all my Navy education had been focused on how things worked. At LMET we talked about how people worked.

We were introduced to Abraham Maslow's "hierarchy of needs." We talked about leadership and teamwork. We did team building exercises and debriefed what we had learned. At first glance some of the exercises were childish but all illustrated a lesson.

We watched the movie *Twelve O'clock High* and studied how Gregory Peck handled each situation. He dealt with each man differently, in the way most conducive to bringing out the best that man had to give. He made them proud to do their part. Atta-boy Greg.

We took a Myers-Briggs Type Indicator. I thought the personality archetype theory was a scam. Since then I've experienced that I could indeed be "typed." If it was a scam, it was a great one.

When I became a supervisor, I could now be more understanding of the individual than the supervisors I had experienced thus far in the Navy. My bosses so far flaunted their authority and chose stepping on their subordinates at every chance. The only tactic I had seen used to motivate people was the threat of punishment. After LMET, I could be different.

School finished, we moved to California taking leave on the way. Our options for taking leave—vacation to civilians—had been limited by schools. Also, we didn't have money to lie by the pool at an exotic resort. We did what all young Navy people did with their leave. We visited family.

Robin and I had been married four years and had lived in Illinois, Florida, and Massachusetts. Now expecting our second child, we set up shop again, this time in California.

In Vallejo we rented a small house on a hillside overlooking the north end of San Francisco Bay. Looking down from the window you could see the Napa River and across the river, Mare Island Naval Shipyard.

A causeway painted light-blue bridged the river to the shipyard. The two-lane causeway carried a steady stream of automobiles on and off the island. Occasionally traffic stopped as the drawbridge rose to let pass a tugboat pushing a barge.

Marine Corps guards controlled access to Mare Island. They inspected each passenger's credentials before allowing the vehicle onto the bridge. Armed and far more serious than the guards I had experienced at school commands, these guys were all business. Impeccably dressed and impeccably professional, they gave off a vibe that suppressed any joking around. While one stood in front of the car looking at Department of Defense decals on the windshield, others nearby watched like a pride of lions hungry and ready.

The Marines' brusque manner gave me a suspicion that, more than making sure my presence here was acceptable, they were actually looking for a reason to jerk me through the window and put me face down on the pavement with a shotgun to the back of my head. Finding my credentials correct and waving me through, I think they were disappointed.

The shipyard was exciting. It was also intimidating. Thousands of people drove on and off the island at every shift change. The place buzzed with activity. Ships and submarines were moored along the riverfront. Tugboats and small watercraft of every description moved along the river or were tied to one of the many piers and docks. There were dry docks and shipbuilding ways. Massive cranes stood like skeleton skyscrapers. Trucks, vans, tractor-trailers, and forklifts were everywhere.

This was not a school command resembling a college campus. This was the real Navy, the Fleet.

USS Nautilus was moored at a pier just south of the causeway. In the process of decommissioning *Nautilus* would soon be towed to New London to become a museum. She was the first nuclear submarine. My submarine, *Seawolf*, was the second.

Seawolf was at sea when I arrived at Mare Island. Each day I reported for work at the Det, a small office building near the submarine berths at the southernmost end of the shipyard. I had heard submariners talk about the squadron their submarine was part of; I suppose my squadron was Submarine Development Group One Detachment Mare Island, even though I never heard it referred to as a squadron. We referred to the organization as DevGroup, and to the building it lived in as The Detachment or The Det.

Submarine Development Group One was the Navy's West Coast home of the Deep Sea Submergence Program (DSSP) headquartered on Point Loma, San Diego. Detachment Mare Island could be thought of as our local office. The other members of our little throng of Special Projects submarines were *Parche* and *Richard B. Russell*. As to the actual work of DevGroup, I knew only whispered rumors and knowing winks.

I asked my temporary boss at the Det, basically a guy charged with keeping me looking busy until my submarine could take me off his hands, what did Submarine Development Group do?

His response was typical of what I was to experience during my entire tenure in Special Projects. He hesitated, more deciding how to answer me without answering the question, which seemed far more complimentary than a professional "that's not your concern Petty Officer Heckman," or the less eloquent but more sailor-like "none of your goddamned business," with which I'd become familiar. I wouldn't have been surprised to hear a call back to the legendary brusque manner of Admiral Hyman G. Rickover: "what the Kindly Old Gentleman decides you should know, Himself will inform you."

I later came to understand that two boats in DevGroup Mare Island had recently suffered some bad luck and had come within an RCH of not returning. Perhaps clandestine service arrogance had subtly shifted to a more respectful approach towards its members. Or perhaps this Navy man, on this day, was feeling generous.

"We are the United States Navy's garbage collectors," he said. His tone closed the subject.

After three weeks of busywork at the Det, I came in one morning to find *Seawolf* returned from sea.

Activity swarmed the boat. Brows were positioned fore and aft and a steady stream of people crossed between boat and pier in both directions. I tried to remember the protocol for boarding a Navy ship. That was boot camp stuff and I had come a few miles since then. Tons of new knowledge lay atop that tidbit of old information. Feeling very small, I watched others board the boat for the correct actions.

You stupid bastard, I thought. The worst that will happen is you do it wrong and get yelled at. You've been yelled at before. No big deal.

Orders in hand, I marched across the forward brow. Stopping short of stepping onto the boat, I came to attention. I saluted the national ensign fluttering at the stern of the submarine and saluted the topside watch. He wore a wide white canvas belt over top a green web belt with a pistol holster hanging from it. The holster looked heavy.

"Petty Officer Heckman reporting for duty. Request permission to come aboard," I said in my most I-know-what-I'm-doing voice.

"Come aboard," said the watch returning my salute. I shoved my orders at him. He glanced at them looking amused. "What division you in?"

"Reactor Controls Division," I said.

"A fucking Nuke." He spat over the side into the green water, then picked up a telephone handset. "I'll call down to fucking Nuke land and see if I can find one of your fucking Nuke pals to come and get you."

Hmmm. Not really the auspicious start I had envisioned.

A minute later a first class petty officer stood before me. "RC Div?" he said.

"That's me." I shook the hand he offered.

"I'm Bob Cranker. I'll be your sea-dad," he said. "Follow me."

Cranker led me along the top of the submarine to the aftermost hatch. I was about to see the inside of a nuclear submarine for the first time. As I leaned over the topside hatch the first thing I noticed was the smell.

16

Racked Out

Lash with seven marlin hitches, being careful that the last two hitches completely close the ends of the hammock over the mattress. Then twist your clews and place them under the hitches.

— *Blue Jackets Manual, 1940 edition*

REACTOR CONTROLS DIVISION Leading Chief Petty Officer, Senior Chief Electronics Technician Johnson, my first LCPO on arrival to *Seawolf*, retired from the Navy. I overheard he and Senior Chief Hicks, the Bull Nuke when I reported aboard and also retiring, chatting about their plans for their next life not to be surrounded by battleship gray paint. The mention of the size of their expected monthly pension check scared me. Too many years have passed since the conversation but I think I recall the figure being more than a thousand dollars less than I was taking home a month. That was in 1985—I was taking home more as an active duty E-6 than they would as retirees. Each man had given 22 years of service. I had a wife, two kids, and a mortgage; there was no way I could keep my household running and gas in the car on their pension check. I mentally filed that information into my "stay in or get out?" folder.

With Senior Chief gone, ET1(SS) Kenneth Kohl assumed the mantel of RCLPO. Kenny had the job nailed down solidly, we were in good hands.

Legions of spines are grateful that hammocks are long gone from the Navy. My time in submarines was the era of bunks with mattresses. When you were not in bed a fireproof bunk cover was zipped closed over the mattress. Fire resistant is probably what the manufacturer's lawyers would want me to say.

Surrounding the bunk was a blue curtain hanging from hooks riding in an aluminum track. Any light-leaks in the curtain were corrected with the clever application of Electric Boat Green, or EB Green, known to the civilian world as duct tape. Within the bunk space was a reading light, an adjustable air-conditioning vent, and a speaker with controls for the ships audio entertainment system. You could select the music channel of your choice and adjust the volume of your personal speaker. That was a nice touch, except that the entertainment system on *Seawolf* did not work.

Lying on my back, the bunk was just long enough for me to stretch out. My feet were at a right angle against the bulkhead at the aft end of the bed as if I was standing on it. My military-short hair brushed the bulkhead at the forward end. I cannot imagine how a man much taller than my six-foot-two could sleep in a standard submarine rack.

Placing my hands together over my chest as if in prayer, one elbow was hard against the outboard bulkhead and the other touched the passageway-side edge of the bunk pan. The turned up edge formed a lip that kept the mattress from sliding off into the passageway. Resting my elbows against the mattress and raising my hands at shoulder height in an inverted push-up position brought my palms flat against the bottom of the rack above me.

My rack was slightly roomier than a coffin. Only slightly. Laying sideways and taking a deep breath, my shoulders wedged against the top bunk. Turning over in bed required a twisting sliding motion with shoulders and hips.

Drawing the blue curtain closed was the submarine equivalent of shutting my bedroom door. Submarine etiquette defined each man's rack as his castle, to remain inviolate. Rack space was personal space to be entered only with permission.

I liked to sleep on my side with a leg pulled up as if I was about to take a stride. I don't know where this habit started; I've done it for as long as I can remember. Perhaps childhood nightmares kept me in constant readiness to run. Maybe my DNA strand has a kink in it that predisposes me to have a preferred sleep position. Whatever the reason, sleeping with one knee protruding at a right angle worked for me.

Arriving to *Seawolf*, the COB game me the choice of available racks in the stern room, the domain of the Nukes. I chose one at waist level, figuring it would be easy to get into and out of. What I did not know until later, was that at sea the ship would often give a jerk as we made a turn or changed speed causing a person to grab hold of something solid to brace himself. He didn't mean to disturb anyone; he was trying to be respectful to the sleeping, a courtesy he hoped would be reciprocated. However, when a man's center of gravity was suddenly thrown past the point of self-recovery, instinct took over. He reached for what was available. He grabbed my knee.

Being jolted out of a wonderfully sinful dream by a rough hand attached to a male voice whispering hoarsely, "Sorry," was a routine occurrence when you have cleverly chosen a waist-high rack and like to sleep with one knee sticking out. If you find yourself in similar circumstances, have a care not to crack your head on the bunk above when it happens.

A nuclear submarine is an amazing machine. More accurately, it is an amazing collection of machinery. Thousands of systems perform specific functions acting in concert to control the ship, provide a life supporting environment for the crew, and conduct the ship's mission.

Space for people on a submarine was a design afterthought. The boat looked as though it were built around the major machinery, with smaller equipment placed in the space between. Cabinets, storage lockers, workbenches, and the like, were tucked into every nook. A three-foot-wide passageway ran down the middle of each compartment, a blue linoleum boulevard. Avenues and alleyways intersected the main passageway, providing access to more equipment. Electrical systems resided in cabinets mounted like tall buildings along each side of the boulevards and avenues.

Steel ladders provided access to lower levels and auxiliary machinery spaces. Hinged hatches in the deck allowed hands to reach valves located underneath.

The submarine made fresh water, cleaned its air, and generated its electricity. Spaces were air-conditioned or heated as necessary. A meal was served every six hours. There were showers, washers and dryers, first-run movies after dinner, ice cream and fresh pies. There was everything a man could want, except sunshine.

Well, everything he could want, if you don't include: skiing at Tahoe, trout fishing the Yellowstone, television, the gym, Starbucks, shopping malls, nightclubs, restaurants, the symphony, mail, current news, magazines that a mother would approve of, live bands, hot tubs, king-sized beds, family, friends, football games on Sunday, the World Series, March Madness, Thanksgiving dinner, the aroma of the Christmas tree, and women. Not included would also be: a walk around the block, seeing farther than thirty feet, barstools, cold beer, gin and tonic with a fresh lime, the wind in the trees, rain on the roof, and women. Your baby girls, your wife or your girlfriend. Rusty the dog, a varying wardrobe, incandescent lighting, and women. Sleeping until you feel like getting up, napping on the couch on a Sunday afternoon, washing the car, and mowing the lawn on a summer day. Nighttime, daytime, fresh fruit, crisp green salad, and milk. And women.

So—not quite everything a man could want was on this boat, but certainly what he needed to survive and perform his duties for an extended period away from the world.

As long as most of the machinery kept working, we could stay at sea until the food ran out. When going to sea for only a few days or weeks, to test out equipment or work the kinks out of the crew,

Seawolf carried a 90-day food supply. Before getting underway for a long mission, groceries were loaded in until there was no more room.

The Navy is awash in ironic terms and wicked twists of phrase. A "working party" is not a pleasant event of music and cocktails attired in business-casual.

"All hands not on watch lay topside for a working party," sounded the 1MC.

On the pier sat a semi-trailer with enough food to stock a grocery store. The crew—that is the enlisted men—formed a line. The line stretched from the truck, along the pier, across the brow to the boat, along the topside deck, and down the hatch. In the boat, the line continued into the galley and the various designated storage locations. Dry stores, refrigerated stores, and frozen stores all had appropriate storage lockers.

The men on the pier and topside the boat formed two parallel lines facing each other. Out of the trailer came cases of canned peaches, crates of fresh lettuce, and five-gallon jugs of milk. Everything moved down the human conveyor. Steaks, roasts, hams and frozen chickens, all down the line of hands and arms, then down the 30-inch diameter hatch into the boat.

Coffee, tea, orange juice, bug juice, potatoes, beans, and rice. Dried pasta, spices, pancake mix, butter, salt, tomato sauce, and peanut butter; all were loaded in with the cooks directing traffic.

It was possible for submarine sailors, known to have tendencies toward anarchy, to artfully position themselves in the line to redirect coveted items. A large can of cashews designated for the chief petty officers quarters was shuttled aft instead of forward and found its way into the engineering spaces. The can was stashed in an out-of-the-way locker.

This is all hearsay, of course. I can neither confirm nor deny anything that may or may not have occurred on the submarine which may or may not be called USS Seawolf in what may or may not be the United States of America, you understand; these are only rumors I pass along to you.

Performing such seditious acts under the watchful eyes of authority figures required boldness of step and steady nerves. The ability to appear to be on important official business was a skill worth developing. Again, so I've been told.

The load-in continued, the standard stowage areas were filled to capacity, yet more room was needed. Institutional sized cans of food were placed on the deck in the berthing spaces; blue linoleum was laid over them. We walked on the temporary "deck." I soon learned to duck my head at the just the right spots even in the dark.

As food was needed in the galley, the cooks took cans away and the temporary deck receded like a fast moving glacier. This was a welcomed event, because food eaten was a measure of time passing. Each can gone from the deck was a day closer to home.

On the seventy-fourth day of a mission, the cooks took away the cans under my rack while I was sleeping. Waking, I rolled out feet first into the darkened passageway. The deck was not where I left it. In the millisecond it took for me to travel the extra fourteen inches down to the real deck, my system pressurized with enough adrenaline for me to have won the Kentucky Derby.

No one slept particularly well at sea. The working areas stayed brightly lit with perpetual artificial daylight. The berthing areas were dimly splashed with red light providing a sense of constant night. The Stern Room where I slept was just aft of the Engine Room. The watertight door separating the two compartments remained latched-open at sea. To help keep as much engine room noise as practical out of the berthing area, a heavy blue curtain was drawn across the passageway two steps aft of the engine room watertight door.

The curtain definitely helped muffle the noise. When the flap was pushed aside, the noise level rose with a crescendo. The sound fell as the curtain swung back into place. A symphony conductor would have envied the consistency of the effect, but it was miserable for sleeping.

As I tried to fall sleep, the constant hum of activity remained on the edge of my awareness. People moving about in the berthing space, showering, clothes washing, the metal door of the head clanging shut, low conversations, and machinery noises from the engine room conspired to keep me awake.

In addition to the persistent noise, the motion of the ship added to the difficult sleeping environment. For the most part, the boat moved smoothly. There was a sensation of motion similar to what you find on an airplane. In the air, the airplane never quite feels solid. You know every minute you are airborne. The submarine was the same. There is no such thing as motionless at sea.

I felt every turn, every up and down, every speed change. It was in the deck, it was in the air, it was in the vibration of the boat. My own heartbeat added to that vibration when the boat moved suddenly. My senses lowered to a quiescent state made sleep possible but did not guarantee its duration. The jerk of a quick depth or speed change roused me from REM sleep to a level of vague awareness. Once the boat settled down, I drifted back into deeper sleep. For a time, anyway.

A man who managed to get several hours of undisturbed sleep was subject to suspicion. When he showed up to relieve the watch in a satisfied stupor he was good-naturedly chided for being a "rack hound." We were jealous.

"Look at you," I said to Mark Gonnella. "You've got pillow marks on your face."

Gonnella grinned. "Almost eight hours of solid rack time," he acknowledged happily. "I slept like a dead parrot."

"'e's not dead," chimed in the Reactor Operator, who was turning over the watch to me. "'e's restin'."

The off-going Throttleman jumped in. "'e's deceased, expired, gone to meet his maker."

The on-coming Throttleman put it to bed. "This parrot is no more!"

The Engineering Officer of the Watch was duty bound. "All right, all right, knock off the levity in Maneuvering," he said. "Secure from Monty Python."

In waters unsafe to be found in, we had an equipment problem that could only be repaired from outside the boat. With the advent of books in the public domain such as *Blind Man's Bluff* and *The Silent War* it shouldn't be any violation of the stack of I-swear-to-never-say-anything-about-this documents I signed on my discharge from Special Projects to reveal to you that *Seawolf* had the capability of sending robotic "Fish" into the deepest depths of the oceans. (I say 'shouldn't' with full awareness that the U.S. Navy is inextricably bound to the U.S. Government, which like all governments, can change, translate, ignore, apply or enforce the rules as it chooses. If you never hear from me again you may rightly conclude that Men In Black have arrived on my doorstep aiming at me that most vicious device of modern weaponry—a subpoena.)

Even in today's wireless era should be no surprise that in the 1980s remote-controlled Fish required tethers and umbilical cords for electricity and hydraulics and cameras and lights and whatever else made them go. (Please recall that I was a Nuke, making the submarine go fast or slow and keeping the lights on was my job; everything I say about Projects gear is rumor and speculation. MIB please take note.) Our equipment problem that could only be repaired from outside the boat involved trouble with the steel cable holding the Fish and its winding mechanism. Repairing it meant surfacing and sending tethered men out to do the job.

Since being seen blows the whole clandestine operations thing, we accomplished the ad hoc procedure like true spooks. We did it at night, in the dark. The boat submerged during the day and idled like a car at an all-day red light. When it was solid dark again, we surfaced and went to work on the problem. The job lasted more than one night.

The sea was boisterous enough to prevent everyone from sleeping during the hours we were surfaced each night. We rocked and rolled in the ceaseless swells of the open ocean. Men were thrown completely out of their bunks. Various contraptions were made out of rope, twine, and Electric Boat Green duct tape looped over piping or other supporting structures to hold men in their bunks. It was a fruitless effort, sleep was impossible.

A weary cheer rose each morning as we submerged into tranquility and waited for darkness. Those on watch during daylight hours were hardest hit. A sleepless night for them was followed by a smooth, uneventful watch spent staring at meters and gauges whose needles didn't move. Indicators blurred and log readings continued down the page unchanged.

The fellows with nighttime watches were not hindered by the roughness on the surface during their normal working hours. They slept soundly while we were submerged.

After three days of this obnoxious routine, the exhausted Reactor Operator in Maneuvering asked the exhausted Reactor Technician out in the engineering spaces to report a parameter. During the seventy-five foot walk to Reactor Compartment Upper Level to obtain the reading the Reactor Tech forgot which parameter he was after. No one fussed at him. The RO repeated the instruction.

The RT wrote the parameter requested on his hand and succeeded on the second try. Innovative thinking to overcome real-life obstacles counts big.

In the Stern Room, just inside the Engine Room watertight door, a deck plate was loose. Each time a foot landed on it, the deck plate clattered. We tried many methods to repair or jury-rig that deck plate into silence. None lasted.

My rack was in the stern room and this clatter irritated me. Trying to sleep, I was roused by each heavy clunk of a foot on the deck plate. It began to wear on me. I was almost asleep, finally, when BANG, the deck plate rattled. Grrr…

A few minutes later, as I was again almost asleep, BANG, the deck plate again. At each rattle I became more enraged. An almost overpowering desire flooded me, a desire to rise up out of my bunk and throttle the man who just stepped onto that damned deck plate.

In the next instant, I realized the effort it would take to complete the act. Exhaustion got the better of me and I gave up the idea. Morality, decency, or fear of prison was not part of my decision; only grim practicality. I didn't want to get out of bed. Murder simply wasn't worth the effort, even a murder sugar-sweet with righteous vengeance.

I don't know whose, but a life was spared.

17

Medical Marvels

Corpsman: short for Hospital Corpsman, an enlisted medic.

— *Naval Terms Dictionary*

MEDICAL ISSUES IN the Navy are handled differently than in the world. In the civilian world when you are too ill to work, you call in sick. This phenomenon of independent action does not occur in the Navy. Calling in sick and staying home in bed will win you a chance to visit with your Commanding Officer. He will talk and you will listen. It will not be a pleasant conversation.

Navy men too ill to work go to work anyway. You may be trained in electronics, radar, sonar, Russian or Chinese, but you have not been equipped by the United States Navy to determine the state of a person's health, even your own. A Navy Hospital Corpsman will determine what treatment you need or will send you to an actual doctor for diagnosis.

The corpsman might send you home to be sick in private. He might also hand you a winning lottery ticket. The odds of either happening are the same.

Regardless of rank, the medical corpsman on a submarine was always "Doc" from the captain to the cooks. Docs were usually interesting characters.

In port I had a small wart growing on a finger. *Seawolf's* corpsman sent me to the Det. The Det, as you will recall, was Submarine Development Group One Detachment Mare Island, the administrative office for our submarine. Among other things, the Det housed the squadron Dive Surgeon. A genuine medical doctor with the rank of Navy captain, the Doc at the Det always had a too-big grin that caused me to suspect a hidden bucket of water poised to dump on my head.

Apparently captains in the Navy Medical Corps march to a different drummer than the average half-crazy submariner. At least this Doc did. Eccentric enough to be interesting, we referred to him as "the Wacko Quacko."

Being a real physician with a real clinic he was just the guy to kill my wart. The treatment involved his laying into the thing with a miniature arc welder. This burned into the wart until I squirmed. That's enough for now, the Dive Surgeon would say. Off I went. I returned the following week for a repeat procedure. This continued for several weeks. The Whacko Quacko and I got to know each other.

He had purchased a deactivated Titan missile silo north of Sacramento and was refurbishing it into underground living quarters. I wondered if electric-arcing a wart to death was an approved medical procedure, or if he had invented it in his missile silo workshop late one night. I did not ask.

I did asked questions about my unsolved medical issues. I had suffered a home improvement mishap with a claw hammer, leaving me with a black thumbnail. Would it grow out or fall off? There was a section of pencil lead broken off in my knee from the seventh grade, should we do anything about it? The Dive Surgeon studied each problem intently. He never played down my issues, treating each one with dignity even though they now seem trivial. I began to feel special. That is, right up until he spied a mole on my abdomen. Low down on my abdomen. It was just above my… well, really low on my abdomen. The Doc looked at it closely. He probed it.

Sticking his head out the door of the examination room he called out to another doctor. He used a phrase no one wants to hear from their physician.

"Hey Joe, come here a sec, you gotta see this!"

Christ, I thought. Doctors have been frowning at that thing all my life and muttering the C-word. Here it comes, the suggestion that I put my affairs in order. After a brief consultation in the hallway, the Dive Surgeon returned with another man in a lab coat. The new doctor studied the mole.

"Son of a bitch," he mused. "Doctor, you are absolutely correct." I cringed, awaiting the dread news. He continued, speaking to the Dive Surgeon as though I wasn't there.

"It's shaped just like Florida!"

Dry-docked at Mare Island for repairs, a barge was pulled in alongside the submarine before the water was pumped out of the dock. This barge was our temporary office, work, and storage space, complete with duty section sleeping quarters. Each division had its own "cage;" a lockable area to safeguard their tools, equipment, manuals, and assorted paraphernalia. Steel-decked and steel-walled, the barge's internal climate was Arctic in winter and Sahara in summer.

Navy personnel did not wear working uniforms off base. Sailors living off-base drove to work in civilian clothes and changed into uniform on base. Submarine sailors whose ship was in dry dock changed clothes on the barge.

Petty Officer Third Class Jim Nuweling was one of the newest men in Reactor Controls Division. We gave him the junior-man jobs. Mostly this meant jobs that everyone else had done often enough to be sick of. Weekly trip point and calibration checks of the Primary Plant Instrumentation System fit snugly into that pigeonhole. Weekly Trip and Cal's were repetitive and monotonous.

The Primary Plant Instrumentation Panel was six feet tall with a drawer for each of the 24 separate Primary Plant parameters measured. The instruments used to simulate test signals were located in the bottom drawer.

The procedure required the technician to alternate between standing to work with the instrument being tested, and squatting to work with the universal test drawer. After two hours of this activity my knees were always creaking. Yes, the Primary Plant Instrumentation trip point and calibration check was a fine job for the junior man.

Jim regularly came to work with a hangover. He was a bachelor and mundane chores such as laundry and housecleaning were not high on his to-do list. When off-duty, he frequented local bars and nightclubs pursuing the fairer sex. On a Sunday duty day he bragged of his successful Friday night safari to the Horse & Cow Bar. By Monday he was uncharacteristically quiet.

Over the next week or so, most of the RC Division men became strangely subdued, going about their work with little conversation. Whatever was bothering Jim Nuweling, it seemed to affect everyone's mood. Not wanting to interfere in another man's private business, we gave each other space.

A few days later, I felt an odd sensation and went to the Doc. Doc Post was friendly, and like all Navy Corpsmen he exuded competence. He quickly learned more about me than I wanted him to know.

My strange sensation was an infestation of pubic lice. Crabs, crotch-crickets, cooties; living things living in my shorts without permission. I left the Doc Shack with a hot neck. How on earth…?

Doc had just run out of the special shampoo needed to kill the little bastards squatting on my property. He said he would bring me some after he made a trip to the Mare Island clinic. Later that afternoon, his resupply complete, the Doc came down to the Reactor Controls Division cage and gave me two bottles of medicinal shampoo for my special problem. One for me and one for a close friend, he explained. I quickly tossed them into my file cabinet clothes drawer.

"Thanks, Doc."

"What's with the file cabinet?" he asked.

I explained the faux locker arrangement. Having seen my drawer at the bottom of the stack, Doc indicated the drawer above mine.

"Who's is this?"

"Mark Jones."

Pointing to the next one up he asked, "And this?"

"Bill Smith," I said.

"So you, and Jones, and Smith, are in the same division," he said with the bemused look of a man who has just seen the light. "Is Petty Officer Nuweling in RC Division, too?"

"Yes, he is."

The Doc says, "Wait; don't tell me." He indicated the topmost drawer directly above Smith's, Jones's, and mine. "Nuweling's drawer must be this one."

"Yes," I said. "How did you guess that?"

Pointing at each drawer from the bottom up he said, "Thursday, Wednesday, Tuesday, and," indicating the top drawer, "Monday."

I didn't know what to make of that. He did say one thing that I understood.

"All the guys who store clothes next to, or below Petty Officer Nuweling's clothes drawer, should wash all their clothes in the hottest water possible. I mean every bit of clothing that has come in contact with anything in these drawers. Your sheets, too. Tonight."

I'm not sure how many wives or girlfriends were lost over the incident. Women have almost no appreciation for a shiny new STD, even a relatively minor one that can be cured with a hot shampooing. Hell, you'd be doing that anyway, right?

"Honey, I really don't know how this happened," floats as well as a man-hole cover. A few days later, round two.

"Sweetheart, about that crabs thing. You're going to find this soooooo funny....there's this file cabinet...one of the guys, whose drawer is right above mine...

Oh, sure.

There were people on the boat that one did not toy with. Obviously there were demarcations of rank and seniority, but there were also key people with the ability to make your life miserable independent of rate or rank. Teasing them or playing jokes on them, even if they were your peers, were very bad ideas.

One never screwed with the cooks, for reasons that should be obvious.

One should never screw with the Yeomen and Personnelmen who worked in the Ship's Office. They administrated your pay and leave. If they asked you to fill out a form for the fourth time, you smiled and said thank you. It was another bad idea to insinuate anything negative toward them for losing the first three forms.

When you saw guys from the Ship's Office out in town you bought them a drink. You wanted them to be your friend. If they didn't like you, well, a typo here, an unfortunate decimal point there, and you were screwed.

Causing a paperwork SNAFU in a massive bureaucracy is easy. Correcting that same problem is not so simple. It may require an Act of Congress, typed in triplicate, each copy notarized by a different left-handed Supreme Court Justice whose birthday falls on February 29th. Not an easy get.

Only a horrible lapse of judgment—or perhaps a loss of the will to live—would cause one to screw with the Doc. There was only one Corpsman on a submarine and he was god. If the Doc decided that your life needed to be more miserable, all he had to do was take the "shot record" out of your medical file and toss it. Of course he would never do that. He was a professional. But he threatened it when he wanted to get your attention. Without documentation of every inoculation received since Day One in the Navy, you would be compelled to get each inoculation again. All thirty or so; including the one with the square needle in the left testicle. It wasn't worth taking the chance that Doc was only kidding about losing your shot record. Death may ride a pale horse, but he can also wear a white lab coat.

As the head and sole representative of the submarine's Medical Department, Doc was also responsible for monitoring the radiation exposure of the crew. Strict limits were set for permissible exposure of Navy personnel that are 1/10th the allowable dose for civilian radiation workers. Each person on the submarine carried a Thermo Luminescent Dosimeter that measured exposure to ionizing radiation. The Engineering Laboratory Technicians read the TLD's monthly and reported the results to the Medical Department, a.k.a. the Doc.

We were in port when a new face joined the *Seawolf* crew. He was a Sonar Technician Third Class, fresh from Sonar Technician School, who thought his new third class chevron carried some weight. It was too bad for him that he had been assigned to a Projects boat. On *Seawolf*, Petty Officer Third Class was the low man on the totem pole.

The new man had a boyish face and cocky demeanor that included a tendency to smart-off at exactly the wrong moment. He did not yet understand that someone acting friendly toward him was not an invitation to make jokes at their expense. He would learn.

One afternoon he managed to say the wrong thing to the Doc. I never heard what was said. As easygoing as the Doc Post was, it must have been quite a stem winder to have riled him.

Doc came aft into Reactor Compartment Upper Level and I was the first person he saw. He asked if I had met the new Sonar Tech. Yes, I had met Young Petty Officer Smart-Ass. He had been wandering around the boat staring at the machinery. He struck me as one of those boneheads who would turn a knob or flip a switch as he wondered "What does this do?" Unauthorized turning of knobs is the fast-track to a short career in submarines. I had given him the nickel tour of the Reactor Compartment to keep an eye on him.

"I'd like to slap both his parents for producing such an arrogant little prick," I said.

"Why do you suppose they gave him to the Navy?" Doc replied. "They probably left him in a basket on the steps of the Pentagon and ran away."

The Doc had a plan. A few minutes later the Ships Announcing System called for the new man to "lay to Reactor Compartment

Upper Level." "Lay" is polite Navy talk for "get your ass there right now."

Arriving in RC upper level, he found the Doc in a worried state. The Doc inquired if Young Petty Officer Smart-Ass had been in the Reactor Compartment earlier in the day. On hearing an affirmative answer, the Doc furrowed his brow even deeper.

"This is bad," Doc said, directing his attention toward me.

"What time were you here?" he asked Smart-Ass.

"About 1030, Doc. Is there a problem?"

Doc turned back to me. "What time did you say you were doing the test?"

Playing my part, I responded, "About 1030."

The Doc groaned. "This is bad."

"What is it?" Young Petty Officer Smart-Ass demanded. He was beginning to worry. The Doc explained, precisely enunciating each word.

"You were in Reactor Compartment Upper Level when Reactor Controls Division was testing the Nuclear Instruments with a live radiation source. To be in the area at that time, you needed to be wearing a special dosimeter, which you weren't."

My role was simple. I nodded gravely.

The story was complete BS, but this kid didn't know that. He didn't know us well enough to suspect we were pulling his chain.

"What kind of radiation?" he demanded. "This TLD is good for gammas and neutrons." The lad had done his homework.

"Yeah, but this different. It's Beta radiation."

"But, Beta particles aren't all that dangerous," claimed our mark. Damn, the kid had definitely done his homework.

"It's Beta...Carotene radiation," I offered. "Beta Carotene, very bad stuff."

Doc shot me a pained look that said: Are you fucking stupid?

It was from some TV commercial or something and had been the first word that came into my head to follow Beta. I wished it hadn't.

Young Petty Officer Smart-Ass was definitely worried now.

"Beta Carotene radiation! I've never heard of that. What does this mean?"

I chimed in. "Jeez, kid, it's been nice knowing you." Doc shot me another look, its meaning clear: Shut the Fuck Up. Originally written with only one line, I was overplaying my part. The Doc led Young Petty Officer Smart-Ass away to work him alone.

The kid was sent to the Mare Island Medical Clinic, at the north end of the shipyard. It was a mile walk in the August sun. A breeze would have been nice, but he had no such luck that day. For all I knew, the Doc may have put in a call and had the breeze killed. Docs can do amazing things.

At the clinic, Young Petty Officer Smart-Ass asked to see a certain Chief Corpsman, who happened to be a pal of Doc's. Coincidentally, the two old buddies had spoken on the telephone just minutes before. Young Petty Officer Smart-Ass was handed a plastic cup with a lid and instructed to produce a semen sample, to be analyzed for exposure to ionizing radiation. Semen, as the story went, was the best source to obtain accurate results in this case.

"Where do I go to get that done?" Young Petty Officer Smart-Ass demanded. Not understanding the collection method, he was expecting a tidy medical procedure. Something scientific and sterile; maybe a cotton swab, or an X-ray. A tube of blood studied under a microscope perhaps. Something with the feel of science to it.

The Chief Corpsman looked at him solemnly. Making a stroking motion with his fist, he pointed to a door with a sign reading "MEN." Young Petty Officer Smart-Ass flushed with embarrassment. Becoming less cocky by the minute, he sputtered.

"But..."

The Chief pointed at the door of the Men's Room. The young sailor was afraid to ask any more questions. When the collection procedure was complete, red-faced Young Petty Officer Smart-Ass handed the warm sample cup to the Chief. The Chief Corpsman handed him a tissue.

The young petty officer began, "No thanks Chief, I already—The Chief Corpsman cut him short.

"For the sweat on your forehead, son. Your results will be in tomorrow, come back in around 1400." A long, hot hike back to the boat.

The next afternoon, a long, hot hike back to the clinic.

"Sorry, son, there was a mix-up with your sample. It got contaminated and we couldn't use it. I need a fresh sample from you." Into the men's room with the plastic cup; the beaded brow, then the long, hot hike back to the boat. Next afternoon; the long, hot hike to the clinic.

"Son, I'm really embarrassed about this. We have a new lab tech and he dropped your sample on the floor. I'm going to need another one from you."

The collection procedure took longer than previously. Young Petty Officer Smart-Ass was rattled by this humiliating jerking-off in a public men's room. A small men's room at that. He stopped each time someone entered for more typical men's room business, so not to make obvious noises. During these interruptions he studied the Uniform Code of Military Justice posted inside the door of every Navy toilet.

Losing his train of thought every time someone entered caused him to lose headway, turning the ordinarily easy task for a young man into a marathon.

He had spent three nights worrying that he was exposed to some horrible mutating radiation. His stress was showing up as reluctance of his manly hydraulics to pressurize properly. When he finally exited the men's room, overheated and out of breath, the Certain Chief Petty Officer had been called away for an actual medical emergency.

Handing the sample cup to the nearest Hospital Corpsman First Class, Young Petty Officer Smart-Ass said simply, "Here it is."

"Here's what?" said the corpsman.

"My sample."

The corpsman held the opaque plastic cup up to the light. "Is this… oh, Jesus Christ, kid what is this for?"

Young Petty Officer Smart-Ass quickly looked around and said in a low voice, "It's my semen sample to test for Beta Carotene radiation exposure." The Corpsman First Class stared at the earnest young sailor. The young man's eyes begged the corpsman to lower his voice. Not a chance.

The Corpsman's voice rising in volume and authority, he snapped the words like a whip. "SAY AGAIN?" His faced darkened; he grew taller. His eyes burned into the hapless petty officer.

"Listen hard, Petty Officer Shit for Brains. If you are fucking with me, I will suture your head to your ass and mail you to the circus!"

That's when the whole thing unraveled. Not in on the joke, the Corpsman First Class spoiled the game. It couldn't have gone on much longer anyway. Young Petty Officer Smart-Ass was growing suspicious and the players in the little drama were close to losing their game faces.

When word got around like it always did, other practical joke aficionados, essentially the entire crew, revered Doc as king.

It was a beauty. The play was simplicity itself, yet subtly complex. An unknown co-conspirator in a neighboring command. Inherent medical and personal secrecy. Three days of suffering. Masterful.

For a week, everywhere he went, Doc was greeted by a patter of applause.

"Doc, you are the Man."

"Well done, Doc."

"Doc, if I have ever done anything to offend you, or for anything I do in the future that might offend you, I sincerely and humbly apologize."

Young Petty Officer Smart-Ass was also greeted by applause, although different in nature.

"Kid, you're buff. Have you been working out? Your right arm is massive."

"Looks like you've lost some weight, son. Don't go ruining your health now."

"Hey kid, help us settle a legal question. We heard you have the UCMJ memorized."

Young Petty Officer Smart-Ass ceased to be. In his place was born Respectful Petty Officer Quiet and Studious. This new guy put his back into it. He was good at his job and was much easier to be around than the other guy.

Never, repeat, never, fuck with the Doc.

18

Drinking Men

It was a woman that drove me to drink
and I never thanked her.

— *Attributed to W.C. Fields*

THE U.S. NAVY did not require alcohol consumption but alcohol was part of Navy culture. The cliché "like a drunken sailor" was not invented out of thin air. Young men and women far from home at a Navy base had limited choices for after-hours activities. Aware that bored young people have a greater tendency to get into trouble, the Navy provided on-base activities at most installations. Movie theaters, TV lounges, and the like filled only a small gap. The place to go was the Enlisted Club or a favorite bar; the things to do was socializing and drinking.

My parents divorced when I was six. My mother focused her life on churchgoing and celibacy. I think she chose God as the one guy who wouldn't disappoint her. She made certain I was at church every time the doors opened, at least three times per week.

The Church of Christ followed the Bible as the inspired word of God, and God would not tolerate us drinking the Devil's alcohol. My closest friends during high school went to the same church and youth group. There were no thoughts of beer, booze, or partying in my crowd. On Monday nights we watched *M.A.S.H.* after our youth group meeting, washing down potato chips and popcorn with Pepsi and Mountain Dew.

Prior to enlisting in the Navy I had only two drinks of alcohol, if they could be called drinks. When I was ten, my Little League team pestered a coach into letting us have a sip of beer after baseball practice one afternoon. I didn't like it.

On a cross-country high school bus trip one of the seniors brought a bottle of Southern Comfort. When the bottle came around to me, I tipped it up like Rooster Cogburn in *True Grit* or so I hoped. When the whiskey hit my throat, I thought I had been tricked into drinking gasoline. Several minutes passed before I could draw a full breath that didn't end in a coughing fit. At the time I found no thrill in alcohol.

When the completion of Basic Training was a week away, my boot camp company was granted weekend liberty. In our dress blues we wandered downtown San Diego in gangs. Most of us were younger than the legal drinking age in California. Word spread that a nearby bar gave more weight to our uniform than to our birth dates. Freshly steeped in respect for authority, we did not want trouble with the law before we graduated boot camp, so we approached tentatively.

"Come on in and have a beer!" the doorman said jovially, like a sideshow barker.

"We're underage."

The doorman became serious. "Kid, if you're old enough to die for your country, you're old enough to have a drink. Come in."

Well, all right then. There must be some legal loophole we didn't know. Adults had been steering me away from bars all my life, I couldn't imagine an adult encouraging me to go inside. Besides, "old enough to die, old enough to drink" was an argument that made sense to me then as it does today. In we went.

The guy at the X-rated movie theater on Broadway said the same thing. By the time the weekend was over I had cast off austerity like an outgrown shirt.

In the early 1980s a member of the Armed Services aged at least eighteen years could drink on base even though the legal drinking age in the state was twenty one.

The Enlisted Club was a popular gathering place. Pool tables, jukeboxes, sometimes a live band, beer, booze, and standard pub grub were available for a bargain price. This was indeed a happy coincidence for young sailors with slim paychecks.

In the mid '80s the rules changed. Military installations were required to honor their host state's legal drinking age. Special dispensation was granted that allowed the Navy to serve reduced-alcohol beer to the underage crowd. Now the eighteen to twenty-one-year-olds, who were still old enough to die for their country, had to drink beer with 3.2 percent alcohol. I was beyond the effected age range by then, but I know it grated on the younger guys. It wasn't so much the watered-down beer as it was the insult.

On the submarine, the Welfare and Recreation Committee subsidized our morale. Mostly this meant drinking excursions. During spring and summer we held family barbeques. A softball game always broke out or at least our version of softball. We played "beerball" with a ball glove on one hand and a beer in the other. The fun was watching a man juggle a softball and a beer can. Dropping the ball brought laughter. Dropping your beer brought catcalls of "alcohol abuse!" Given the choice, we let the ball whiz past untouched to avoid spilling our beer.

In 1980, Candy Lightner got serious about stopping the adverse combination of booze and automobiles. She founded Mothers Against Drunk Driving and changed the world. As a result of her efforts the Federal Omnibus Drug Enforcement Education and Control Act of 1986 became law. Although the clouds of public opinion had been steadily darkening over the heads of those who climbed behind the wheel impaired, the storm now raged with full force. Roadway checkpoints became the norm. Those caught over the limit went to jail and felt a lasting sting deep in their pockets.

In addition to heavy fines, attorney fees, and court costs, a drinking man's job came into jeopardy if he got a DUI. A Navy man with a security clearance might find that clearance revoked, and orders with his name on them to a skimmer in the Indian Ocean, a submariner's vision of purgatory.

Cautious of "double jeopardy," the Navy would not prosecute a person for an offense he was already on the hook for with civilian authorities. Those troubles were considered personal. Personal troubles were something to be left outside the fence.

Instead, the Navy prosecuted the unfortunate person for a myriad of related offenses, such as dereliction of duty, conduct unbecoming, disobeying a direct order (driving under the influence was contrary to a standing order), and of course, his security clearance could go away. If he was incarcerated in a civilian jail so that he was unable to report for duty the next morning, he was guilty of Unauthorized Absence. If his ship left without him, he was guilty of Missing Ship's Movement.

Enlisted military personnel are subject to Non-Judicial Punishment. NJP, referred to as "Captain's Mast," was an informal courtroom style proceeding held by the commanding officer. Rules of evidence did not apply. Sole discretion was given the captain to mete out what retribution he saw fit.

Captain's Mast could trace its heritage to sailing ships and floggings. In the modern Navy, the guilty were fined, restricted to the ship, reduced in rate, or rehabilitated in whatever manner the CO deemed appropriate in accordance with the Uniform Code of Military Justice. Captain's Mast was the Navy parallel of a parent punishing a rebellious teenager.

A Navy man with a heavy alcohol habit courted disaster.

Under the heading of risk management, the Navy decided that sponsoring or sanctioning events serving alcohol was poor policy. Welfare and Rec continued to provide soft drinks at the BBQ, but now you had to bring your own beer. The command discouraged the keggers that had been routine.

The command wanted notice of all planned parties and held a conversation with the host prior to party time. If anyone was cited for an alcohol related issue because of the party, the host would be held accountable.

This felt like a cheap shot, command and control at its worst. Throwing a party became professionally hazardous.

With this new intolerance an old submarine tradition was officially laid to rest. In the bad old days a newly qualified submariner would find his coveted dolphins at the bottom of a pitcher filled with booze. Vodka, gin, bourbon, scotch, rum and whatever else could be found, were all dumped in together. The nastier the mix the better we liked it. For him, you understand—the other guy. He drank until the metal emblem caught in his teeth. He was then a whole man and a submariner.

Sure, here and there a man died from alcohol poisoning, but clearly he was not submariner material anyway and therefore was of no concern to us. In the mid '80s "drinking your dolphins" was outlawed. Like most things outlawed, the practice simply moved underground.

I'm not certain of the motivation, but the Navy took on the view that instead of simply booting addicts out of the Navy as was the practice previously, the thing to do was treat the offender.

Business and industry were under pressure to adopt drug and alcohol treatment programs, as well as provide other services to employees. Social pressure of the times became politically expedient. More likely, perhaps the cost to replace a trained Navy person compared to the cost of rehabilitating him worked out in favor of the latter.

Whatever the reasons, because my job description carried the word "supervisor," I attended a one-week training course to learn how to recognize a substance abuse problem in one of my men. Or perhaps in my own shaving mirror.

All supervisory personnel cycled through this program and we looked forward to it. A week off the boat was a week off the boat and the school was held from eight in the morning to four in the afternoon. "Drinking School" was a vacation.

On Monday, before the first class began, I loitered with students from other commands on Mare Island. We planned to meet at the Top Four Club for lunch. Burgers and beer sounded perfect for a vacation afternoon. Imagine our dismay when the instructor informed us that we were not to drink during the course of the training, even after hours. Sure, buddy, say what you have to say. We were still going to the Top Four.

Near the end of our first day of class, we took a survey similar to the survey used to determine if one of our subordinates had a substance abuse problem. This would familiarize us with what a person with a drinking problem would experience in the real class for drunks. The survey covered other drugs as well; cocaine, marijuana, amphetamines, etc. Having no desire for those I checked their boxes no.

The instructor left the room while we took the survey. He explained that it was for our own use and would not be turned in. There were no incorrect answers. Fine. As soon as he left, the mood lightened. We helped each other with what seemed like trivial questions.

A question asked about our daily alcohol consumption in terms of ounces. Who the fuck drank by the ounce?

"How many ounces in a can of beer?" someone asked.

"Twelve."

"Sixteen."

"No, that's a tall-boy. A regular can of beer is twelve ounces. Have you been riding the Schlitz Malt Liquor Bull, cowboy?"

"Don't diss my Schlitz, twidget. It's got more ooomph than that sex-in-a-canoe shit you drink."

"Sex in a canoe? What are you talking about?"

"Your light beer is fucking close to water."

I checked yes next to the question, do you drink more than eight ounces of alcohol at one sitting. That must have been a trick question, I reasoned. No one would drink only half a beer. Usually I didn't notice the first six beers having any effect.

Another question asked if I ever drank by myself. Only when no one else is around, I thought. Another yes check mark.

The survey asked if I ever planned my day around drinking. Well, we had just planned lunch with beer at the club. Yes.

Did I ever feel bad about my drinking? On a hung-over Sunday morning, hell yes, I felt bad about my drinking. One time in Waikiki, when Jim and Rob and I had decided that tequila was the right thing to do...never mind, that's not important. I checked the yes box.

"According to this, all my guys are drunks," declared an LPO from another boat. A guffaw rattled around the room. He was kidding, of course.

"Yeah," said another sadly, "none of my guys can stay sober long enough to take this survey." He seemed sincere.

"Now they want to take away our booze," lamented another LPO. "Next thing ya know they won't let me touch myself in that special way."

His friend jumped in, "Not at morning quarters anyway, Chubbs!" The men from their boat laughed. Apparently this inside joke had a back story. I don't remember if Chubbs was the man's real name or a nickname. It's probably best we don't know.

The survey had painted us all as alcohol dependent. Damn, that was depressing. We considered ourselves as alcohol sustained. We weren't involuntarily trapped under the wicked spell of that devil rum; we went willingly, gratefully, gleefully. Until there could be an unwind button installed in submariners, alcohol would serve. Right or wrong, partying was our psychological equivalent of a hard drive defrag program.

The day's substance abuse schooling over, we headed to the Horse & Cow to ponder our moral deficiencies over a few beers.

19

Endless Training

The more you sweat in peace, the less you bleed in war.

— *Chinese proverb*

A TYPICAL FAST attack submarine organizational chart included four departments; Weapons, Operations and Navigation, Supply and Medical, and Engineering. *Seawolf* added another department; Projects.

Departments were broken down into Divisions, according to specific skills and areas of responsibility. Every job as well as every piece of equipment or materiel was assigned to a particular division.

Most divisions were named for the major area of their responsibilities. Radiomen (RM), Sonar Technicians (ST), Electronics Technicians (ET), cooks (Mess Specialists, MS), Torpedomen, Auxiliary Division (A-Gang), Quartermasters (QM), and so on.

The Engineering Department consisted of Electrical Division (E-DIV), Reactor Controls Division (RC-DIV), and Machinery Division (M-DIV) with Engineering Laboratory Technicians (RL-DIV, a.k.a.- the ELTs) as a subset of M-DIV.

Today I hear civilian friends talking about getting training in some new facet of their job. This often entails attending a class held in a hotel meeting room. The quality of the class usually coincides with the quality of the hotel.

Once completed, the attendee is issued a certificate printed on a stock blank from the office supply store with a gold foil star of excellence. The certificate states: "This person has completed a course in Lean Manufacturing," or in "Political Correctness in Human Resource Management," or in whatever the topic. Now a civilian, I possess several of these.

After this training, the attendee goes back to work with another positive check mark in his employee file, and perhaps an increase in seniority or a new aspect to his job. This person is said to be trained in the topic. This person is now considered an expert on the subject.

In the Navy, a school or course taken may have this sense of beginning and end, but learning in the discipline does not stop. Training is not a thing to be over and done with. In the Navy, training is a way of life.

Once a man was schooled in his rating—his subject of expertise—he still faced plenty of hurdles in the name of professional development.

Departmental training was held each week. Engineering department training covered items like casualty procedures, radiation exposure requirements, reactor safety precautions, and propulsion plant maneuvers. Specific topics that engineering department training could be held on were endless. Radiation exposure prevention, radioactive contamination prevention, procedures to "free release" an object suspected of being contaminated, and the like, would be prepared and given by a member of ELT division.

Steam turbine theory might be given by a Mechanical Division member. The details of operation of 450-volt AC circuit breakers may be given by a member of Electrical Division. Reactor Controls gave departmental training on reactor control systems and operations. A typical topic might be the progression of ones and zeros through a logic circuit.

The Medical Department, the Doc, gave Engineering Department Training on first-aide. The Navy was serious about cross-training. Everybody on the submarine needed to know everything, keel to periscope, bow to stern.

Engineering Department training included conducting casualty drills. Drill sets were debriefed afterward for everyone's learning. Correct actions taken by watchstanders were reinforced. Incorrect actions were discouraged. Doing too much of the wrong thing could get a man disqualified from a watchstation. He would have to re-qualify like a new guy.

When underway on a real mission—that is to say a mission that if discovered, another nation would seek to kill us—we were cautious about conducting drills. *Seawolf* was an old boat. Causing unnecessary plant transients was laughing in the face of Murphy's Law. Going boldly where no man has gone before didn't mean we had to go stupidly.

In addition to departmental training, divisional training was held weekly. Divisional training was more specific to the needs and knowledge set of the particular division. Preparing and delivering training was a load shared among the division's members. Preparing and delivering training to one's peers is itself a training method.

Each piece of divisional machinery or equipment had a specific person assigned as the system expert. When assigned as a system expert you learned everything there was to know about the machinery in your charge. Theory of Operation, Technical Specifications, what it was made of and why, troubleshooting techniques, how to find spare parts, and on, and on. You also became expert with the paperwork. Detailed maintenance records were kept on everything in the boat since the ship had been commissioned in 1957.

A Nuke newly reporting onboard for assignment to the 'Wolf was expected to qualify on all of the nuclear watchstations normally associated with his rate. His own desire to excel might cause him to also qualify for watchstations outside his rate. Also on his to-do list were advanced watchstations, for example, Engineering Watch Supervisor.

The other members of my division wanted me qualified ASAP to ease the watchstanding and maintenance burdens of the already qualified. Every once in a while a person dragged his feet on Nuke quals. Reading the Sanskrit on the bulkhead, a smart guy knew that once nuclear qualified for this reactor and propulsion plant his personal workload was going up. Free time would go down proportionately.

Instead of getting qualified as quickly as possible, an individual sometimes paced himself. This person was derided as a "skate," a "skating motherfucker," a "sandbagging motherfucker," a "dink motherfucker," and sometimes just plain old "motherfucker."

The motivational tools of the trade to help people become qualified were primarily threats of never experiencing happiness until qualification was obtained. Also useful were heaping portions of ridicule, guilt, and shame. Having been raised in a home heavy on religion, I was well-acquainted with guilt and shame. It was ridicule that pissed me off.

Men with more time on the boat were willing to help me and the other non-quals learn. Even so, they resented a bullshit artist, a brown-noser, or a goldbrick, especially all in the same person. No one likes a salesman: that guy who works hard to sell you on how smart he is. Instead of talking to the man himself, you're talking to his publicist. Instead of biding his time until he naturally becomes one of the gang he tries to crowbar his way into acceptance. That tactic hasn't worked since Abel told Cain, 'you know I'm dad's favorite.' Most new men eventually quit trying to impress and turned out to be fine guys.

Another archetype was the new guy who came from another submarine. Invariably he would regale us with how well his last boat handled some issue that he thought we handled poorly. It was usually a matter of "differently" instead of "better or worse."

"Back on the (insert former submarine name here) we did it thus and so…" the new man would say. The reply to this was always well-argued and inspirational in order to encourage the new man to think more broadly.

"Fuck you and fuck the (insert former submarine name here). You're on *Seawolf* now. THIS is how we do it on *Seawolf*."

When asked a qualification question in a system checkout and not knowing the answer, the Usta-Boat guy would spout off how the system in question worked on his last boat.

"Well, you know, back on (insert submarine name here) that system was pressurized to 1300 psi and the isolation valve was right above the coffee pot," he might say. The sailor giving the checkout would respond predictably.

"Who gives a shit how it worked there? How does it work here?"

The Usta-Boat guys always got grief until they relaxed about their past and focused on their present. Their experience on another boat was valuable, yes; no one liked it as an excuse. Tell me what you know, not what you used-to know.

20

Practical Factors

Never wrestle with a pig. It annoys the pig and you get dirty.

— *Attributed to many*

A NEW MAN was also expected to "qualify on the ship." Qualifying "Ship's" was done simultaneously with his nuclear quals, but qualifying on "the boat" took a backseat to Nuke quals. Although important, required, and desired, qualifying on the ship was treated by Nukes in much the same way Freshman English was thought of by an undergrad majoring in Astrophysics. Passing the course was a given, but it was sure to be a pain in the ass.

Ship's qualification was about the boat. How the submarine worked, where important things were located, how they operated, and of course, why. What level of catastrophe could we recover from, and what must be done to ensure that? What troubles were we less likely to recover from?

When men working toward qualification did not turn-in the required points each week, they were placed on the Delinquent list; publicly posted. Once "dink," life got miserable until you redeemed yourself by catching up your qual points and getting off the list. I mean, you know...that was the rumor.

When I first came to *Seawolf* from the S1C prototype, I officially signed my name as "Karl Heckman ET2." A month later I was advanced to ET1. Now I was "ET1 Heckman." The advancement in pay grade put more dollars in my paychecks but added only a nickel's worth of seniority to my resume.

Each step up the pay grade ladder I made moved me into a different club. Now I was a first-class and other first classes treated me as one of them. I was now entitled to the Top Four Club on Mare Island. The "Top Four" was reserved for the top four enlisted ranks—E6 through E9—in all branches of the military.

The Navy made a valiant attempt to add prestige to advancement. Being welcomed into a new club both professionally and recreationally was a perk that added a subtle sense of increased self worth. Happily for the taxpayer, that benefit was provided without added cost. However, the thrill of my acceptance into a more senior group diminished quickly on finding that my job did not change one iota.

Even though newly advanced in rate, I was professionally incomplete. I lacked the designation of being submarine qualified. Without dolphins I was not much more than a visitor on the boat, a perpetual new guy. Until submarine qualified, I was an outsider, a party crasher at a family reunion.

The desire to sign my name "ET1(SS) Heckman" and be one of the crew—to really be one of the crew—was a force more powerful than any commandment to get qualified.

Much of Ship's quals was made easier by the boat being in dry dock during my early days on *Seawolf*. I could lay eyes on things other sailors working on their quals at sea could only see in drawings and pictures. I could also see equipment under the most massive tarp I'd ever seen that gave clues to the special capabilities of our submarine.

Our dry dock time also meant I had to wait to experience the boat in operation. Many of the Practical Factors—Prac Facs—that required me to operate ship's equipment underway also had to wait.

One practical factor I completed at sea was a rite of passage on *Seawolf*. Locating and breathing from all the emergency air manifolds on the boat was a necessary skill. These manifolds had quick-disconnect fittings to accept the corresponding fitting on an Emergency Air Breathing mask.

An EAB mask had a clear plastic face plate and sealed tight to your head, like the kind civilian firemen wear. The rubber straps that tightened it to your face usually pulled out hair when removing the damned thing. A tight seal to your face was important, of course, to keep out all the bad things that might be in the air outside the mask. Good air came into the mask through a fifteen-foot-long rubber hose with a quick-disconnect fitting to match the Emergency Air Breathing manifold.

The emergency air manifold fittings were easy enough to operate with two hands, but a man at a fire scene or in the middle of an air hazard would have his hands full. A professional needed to operate the EAB fitting one-handed.

Operating the fitting with one hand required dexterity. The same hand that held the hose and its fitting had to push that fitting into its mate while simultaneously pulling a spring-loaded locking ring toward you. Imagine plugging a portable electric drill motor into an electrical extension cord using only one hand. That's the motion. Practice helps.

Operating the quick-disconnect fitting one-handed, and under pressure, took both dexterity and presence of mind. Air did not flow into the mask until the fittings were properly mated.

The standard test to determine if a man knew the locations of all the manifolds and could competently operate the EAB mask and its fitting was simple. Put the man in a mask, seal it tight, and send him from one end of the boat to the other. The EAB manifolds were often more than a hose-length apart. The man would have to disconnect from one air source and find the next while holding his breath.

The task itself was simple in theory. But the level of difficulty of the practical factor as I have described it was far too low for submariners. After all, the conditions that would necessitate EAB masks would likely be uncomfortable and hazardous. The possibility of smoke obscuring the vision was real, inoperative compartment lighting was likely; the path hindered by debris and people was also a likely scenario. To significantly raise the challenge of the test only required simulating what could be real conditions.

As I prepared to perform this practical factor I was handed an EAB mask. This mask had been specially prepared for my test. The clear plastic face shield had been taped over with a piece of black plastic trash bag. Through it I had a view of the murkiest sort. I could see only vague outline shapes in front of me and no detail whatsoever.

Way back in boot camp, during firefighting training I had experienced the "fire room." A concrete structure; stacks of wooden shipping pallets set ablaze inside. Recruits went in, holding the belt of the man in front of them. We found out firsthand what a fire in a confined space was like. The choking smoke and the blast of heat had stretched our undeveloped nerves to the limit and classroom knowledge fled our brains. Nearly every man had to be reminded to get low, stay calm, and conduct himself as he had been trained. Inside the "burning building" experienced instructors oversaw the exercise. But we rookies didn't grasp the significance of that fact. The physical shock of the situation had been so great that we didn't notice that men, who did as they were telling us to do, were surviving the ordeal in fine style. The instructors were coolly doling out instruction and encouragement. Here and there a recruit's self-discipline gave way and he bolted for the great outdoors.

When that taped-over mask came down over my head, the memory of the fire room came back in a flash. Claustrophobia wasn't in my nature, but I did enjoy a shock of panic as the straps tightened the mask around my face. Of course, I choked it down and steeled my nerves, as would any qualified submariner. I tried to talk down my rising heart rate.

A moment later another surge of adrenaline shot through me. A palm blocked the end of my air hose. The mask collapsed into my face as I labored to breathe like a goldfish flip-flopping on the floor. The hand was removed and I sucked in air.

"Motherfucker!" I said a deep breath later. "Warn me when you're going to do that." A voice in the dark chuckled.

"Just testing the seal, non-qual, don't get excited," the voice said.

I started the EAB practical factor as far aft as a man could get in *Seawolf*. A qualified submariner accompanied me on the trip for my safety and his mirth. It was a little like pin the tail on the donkey. My monitor kept me from bumping into and snagging things that would kill me or cause trouble for the boat. Scrapes and bruises on my person didn't bother him.

Making my way through Engine Room Upper Level, I had to disconnect my air and pass into Reactor Compartment Upper Level. It was long way to go without a breath. Disconnecting the air fitting, I turned and ran straight into a roadblock.

"Excuse me, sir," said a happy voice in front of me. I moved to my right. The body facing me moved to my right. Blocked again.

"Oh, I'm soooo sorry," said the happy voice. "Please, pardon me." As I moved to the other side of the passageway the body moved to again block me, but I grabbed it by the shoulders and whisked around it like a defensive lineman rushing the quarterback. I had no air to waste on courtesies or insults.

Ducking through the watertight door into the reactor compartment laughter and applause followed me. "Atta-boy, blue-shirt, don't take any shit from the Chief!" I scarcely heard. All my being was focused on the next air manifold. My calm, cool demeanor was close to black-out, the world was going fuzzy on me. My fingers fumbled with the air fitting. It snapped home and cool wonderful air filled my lungs. I stayed connected to that manifold until my chest stopped heaving.

That was close. I had come within a heartbeat of ripping off the mask and sucking air to avoid blacking out. But I continued on.

The Torpedo Room was crowded and the Torpedomen messed with me too, but by then I was wise to the game.

21

Visiting Brass

The better part of valor is discretion.

— William Shakespeare, King Henry IV part I

SHIPS FASCINATE ME. A Man-O-War is exciting to a young man the way the rumble of a V-8 engine harmonizes with his soul. A warship's dormant weapons imply power, and caution those who would challenge the ship and her crew. Even at rest, the smooth curve of a warship's hull promises speed, horizons gained, and far-off destinations. Tied to the pier the ship waits for the mundane business of people to finish, so it can once again return to the big water and continue the epic of men and the sea.

Poetry aside, submariners—always irreverent and unimpressed by majesty, tradition, or daunting armament—refer to all surface vessels as "targets."

Leaving Mare Island to put to sea, our first stop was a few miles down San Francisco Bay at Naval Air Station Alameda to load weapons. I never figured out why an air station was the place to go for submarine torpedoes. I suppose torpedoes are basically bombs with propellers.

Returning home we again stopped at Alameda to return the weapons. I was never involved in weapons handling in any way but apparently we checked out torpedoes from a weapons lending library.

Once at sea we severed the umbilical cord of mother Navy. This phenomenon contributed to the independent spirit of submariners and also to the high standards of professional competence they demanded. We were on our own. Whatever situation presented itself, it would be up to us and the resources we carried to complete the mission and get ourselves home. Everyone had to pull their weight and a little more. There was no tolerance for slackers or incompetents.

It was common to find a man reading an equipment technical manual simply because he wanted to know. In submarine culture, knowledge and correct actions under duress are respected above all else.

The boat carried machinery to make water and electricity and to clean the air. The reactor carried fifteen years supply of nuclear fuel contained in a core structure not much larger than a single barrel of oil. The fissionable material in the core assembly is arranged to produce an even and controllable power distribution throughout the core. If one could suspend the physics of fission and the mechanics of fuel cell design and empty all the fuel out of the reactor it wouldn't fill a five-gallon bucket. That's an astonishing amount of power to be drawn out of a little pile of dirt.

Provided the machinery continued working; the limiting factor on the length of a voyage was keeping the crew alive.

Groceries we brought with us, fresh water we made. The 8000 gallon per day evaporator used 150-psi auxiliary steam as a heat source to distill seawater into potable water. We referred to the machine as "the 8K," or as the "Evap-a-tron" in jest. Its WWII-era heritage juxtaposed sharply with the modern electronic Navy. We always spoke of it respectfully to avoid offending the water god and bringing grief to us mortals.

The ability to make fresh water was critical for a nuclear powered vessel. Beyond the obvious needs of the crew, the steam plant required water to produce power. Makeup-water must be added to replace the losses of valve stem leakage, pump shaft leakage, and other normal sources of water loss. Without replacing the losses the steam plant can operate only until the water levels drop dangerously low. We would then be compelled to shut down the reactor to prevent damage to the steam system.

Without steam, turbines cannot turn electrical generators or propellers. Without electricity the submarine ceases to live and is not much more than a billion dollars worth of spare parts bolted together. To keep steaming, we had to make fresh water.

The Evap-a-tron was cantankerous. It broke down episodically, but usually was back up and making good water within a few hours. But not always. When water making went down, water usage was instantly curtailed. Clothes washers were stopped mid-cycle. Showers were secured, no matter who was lathered up. The cooks broke out paper plates and cups. The only people allowed to shower when the evaporator went down were the cooks. No one complained about that.

During these periods of reduced water usage, baby powder was the submarine sailors' friend. A few sprinkles of powder freshened bed sheets and body. The crew took on a sweet fragrance.

We were headed home but still two weeks out when the main water distilling plant went down for the count. After a few hours of restricted water use I quizzed Machinist Mate John Uran.

"John, I gotta wash some clothes, what's with the 8K?"

"The Evap-a-tron is tits-up. All four paws in the air and Xs over its cute little eyes." He mimicked the scene in case I couldn't picture it myself.

Soon all off-watch Machinery Division personnel were knee deep in the aft bilge bay in Engine Room Lower Level trying to revive the beast.

The 8K was difficult to work on. Corrosion caused by hot brine and steam, made the work dirty, hot, and miserable. Rusted bolts broke and had to be drilled out. Rusted nuts rounded off in the wrench and had to be cut off. The 8k's location was inconvenient, half beneath the deck plates and tight against a bulkhead. Skinned knuckles and shortened tempers became an epidemic. While M-Division sweated, the rest of us griped.

The Engineering Officer of the Watch calculated how long our make-up water supply would last. Keith Lyly started calculating how long it would take to get home on the emergency diesels. Neither calculation produced happy news.

The two General Motors diesel engines mounted outboard on either side of Engine Room Upper Level drove generators to produce electricity. They could charge the ships battery which was our primary backup system of power generation and propulsion. The diesels could make enough electricity to keep vital equipment operating and to power the Emergency Propulsion Motor. However, the generating capacity of the Diesels was a fraction of the power our S2Wa reactor plant could produce.

The Emergency Propulsion Motor would not turn the propeller shafts as fast as the Main Engines. It would be a slow trip home with pitiful few lights on. We were also concerned that *Seawolf's* top speed on the Emergency Propulsion Motor would be unable to overcome the tide running out of the Golden Gate.

Forty percent of California's runoff water drains into San Francisco Bay. The Bay is filled and emptied by the ocean tide. Other than evaporation the only outlet for all that water is Golden Gate Strait. The strait, carved out by a glacier long gone, is only twelve hundred yards wide and three hundred feet deep. The power of the water rushing out of the bay through that funnel is immense, even at slack tide. If we couldn't get through it with our reduced power, we would at least be close enough for a tug to come out and tow us in. That would be embarrassing, but survivable.

Every system on a submarine has a backup. Even though it was much smaller, our backup electric ion-exchange distilling machine was making good water. With that makeup water, and by tightening down sources of water loss, the steam plant kept operating.

On a sunny afternoon we steamed into San Francisco Bay on nuclear power.

M-DIV had worked continuously on the 8K for a week and was still working as we tied up at Alameda to off-load our weapons before heading home to Mare Island. We moored to a pier as wide as a football field with a parking lot in its middle and the aircraft carrier USS Carl Vinson was moored to its opposite side.

Seawolf's weather deck, where we stood topside, was six feet above the waterline. Vinson towered over us twenty stories. At one-fifth of a mile long it blanked out the view. Conditioned to the fluorescent lighting in the sub, our eyes were sensitive to sunlight. We held up our hands to shade the glare and squinted to see the top of the carrier.

The Machinery Division officer, properly the Main Propulsion Assistant, or MPA, was Lieutenant Murray Snyder. Lieutenant Snyder had spent the last several days in the thick of it with his men wrestling with the evaporator. Even though he was supervising the job instead of twisting wrenches, being in proximity of the work covered him in rust and grime. He handed the working men tools and parts without hesitation. Weeks at sea and working head-down in the bilge for the past week without a shower had beaten him down. Tired and filthy, he trudged toward the vending machines at the head of the pier. Every man on the boat was familiar with the vending machines on the pier at Alameda. Men had started counting their change in anticipation two days earlier.

Day dreaming about the California sunshine and a frosty Coca-Cola had kept Lieutenant Snyder going for the past week. Now he had the former and was on his way to the latter.

Two aviation officers from Vinson were loading the trunk of a Mercedes convertible. Their white uniforms were dazzling, gold braid shone in the sun. Their shoulder boards designated them as commanders. The sight of our lieutenant in his dirty khaki's sporting a five-day beard was too much for their professional pride to bear. They berated him as a disgrace to naval officers everywhere and demanded to see his superior. Lieutenant Snyder pointed toward *Seawolf* and her Captain monitoring the torpedo unloading.

That wasn't good enough for the pilots; they ordered the Lieutenant to lead the way.

Commanding Officer USS Seawolf was a resolute man of few words and the embodiment of command presence. A bushy red mustache characterized him as an independent spirit. The crew admired that. His uniform creases were the sharpest on the ship. Given the challenges of space and the lack of laundry facilities on a submarine, I never understood how he accomplished such an impeccable appearance. Instead of talking about uniform smartness, he led by example. The crew admired that, too.

We respected him as our skipper and as a man. Had our captain wanted the fires of Hell extinguished with a bucket of snowballs, we would have said "aye aye, sir!" and done the job.

Seawolf's captain politely thanked the two Vinson officers for their concern and assured them that he would deal with Mr. Snyder's appearance issues. That should have ended it, but the pilots wouldn't let it rest. They insisted on knowing what brand of military justice the Captain was planning for the Lieutenant.

The Captain had already spoken more words than he preferred and he certainly felt no obligation to oblige interlopers regardless of rank. He bid the aviators good day and indicated the brow leading to the pier. Unaccustomed to being dismissed, the aviators hesitated. They preferred being revered. Red-faced, the senior man protested.

"Sir, I am the Commander of Air Wing— " He never completed his complaint.

"Topside Watch," the Captain barked. A Torpedoman Third Class popped to attention beside the skipper.

"Escort these gentlemen off my submarine." Expecting more esteem for their rank and position, the polished aviators were stunned.

The air wing commander's companion blustered, "Who in hell do you think— "

His jaw went slack in mid sentence. *Seawolf's* skipper and the bustle of activity topside were suddenly and completely eclipsed by the distinctive ring of the topside watch racking a round into the chamber of his Model 1911 .45 caliber automatic.

The pistol pointed skyward almost casually, as a conductor might ready an orchestra for the opening salvo of a symphony.

The offended officer forgot whatever he was about to say. The topside watch's eyes gleamed. He had become the focal point of the Vinson officers' universe. He was ready. His jaw set, his brain screamed; oh please, oh please, let me shoot these bastards…

Ashen-faced the aviation officers marched across the brow and out of sight. The Captain turned to Lieutenant Snyder.

"Lieutenant, what the hell did you do to those guys?"

"Captain, I swear, I was just walking past them on my way to get a Coke from the machine up the pier. I saluted and said, 'Good afternoon.' They thought my appearance was unmilitary."

"No shit," the Captain replied. "You look like hell."

"Yes, sir," said the lieutenant, unashamed. "I've been working on the evaporator."

"I hope a can of soda was worth the trouble," said the Captain.

"I never got that far," Mr. Snyder said. The CO dug into his pocket for a handful of change.

"Let those Airedale assholes get out of sight before you go. Bring me a Dr. Pepper."

Within the hour the entire boat had the scoop on the interchange. One of our shipmates had been wronged; even if he was only a lieutenant. He was our lieutenant, by god, and a Nuke at that. If we Nukes wanted him abused we would do it ourselves, thank you very much. Throughout the boat, grumbling rumbled about the goddamned surface fleet and fly-boy sons-of-bitches.

Later I discovered that even though they sailed on the same vessel, the sailors of the ship and the aviation community on board were different entities. That distinction was lost on us. We lumped all non-submarine organizations together as "skimmer pukes" and "targets" and as such included them in the long list of people we would not invite over to the house.

Electrician's Mate First Class Keith Lyly and I were enjoying the sunshine topside on the ship's stern with a couple of forward guys whose names and faces I am unable to recollect. (Nothing personal guys, if you are reading this, I just don't recall who was present on this occasion. If you were there, give a shout.) Keith and I were pals, even away from the boat. We both liked hunting and shooting, and along with our like-minded shipmate, Machinist Mate Rob Kersch, we loaded the most precise rifle cartridges we could in my house in Fairfield. Keith and Rob were single men and the little workshop in my garage became a sanctuary for the three of us. We loaded our ammo and puttered with our rifles. Our quest was supreme accuracy and we went to whatever lengths necessary to wring the absolute best out of our bolt action hunting rifles. We then went to the shooting range and tested our work. Then back to the workshop to evaluate and prepare rifle and ammunition for the next testing cycle. At sea we traded hunting and shooting magazines with favorite articles dog-eared.

On this afternoon we talked about the wrong inflicted upon our Lieutenant by the fly boy chumps. Gruesome medieval retributions were bantered about until a halfway plausible idea finally took shape out of the roiling thundercloud of youthful indignation. Plans were organized and the group purposefully dispersed.

Seawolf was fitted with two signal ejectors; one forward, one aft. These used 400-psi compressed air to launch flares and other signaling devices while submerged. Fixed in position as an integral part of the ship, their aim was not adjustable. The aft signal ejector pointed 15 degrees from vertical in the direction of the USS Carl Vinson. Perfect.

Searching for suitable projectiles, we discovered that the bore diameter of the signal ejector was conveniently sized to accept a large orange. The first firing attempts were bloopers into the water between the boat and the pier. The fruit was too small to seal the air pressure. Larger fruit jammed in too tightly and disintegrated into a spray of Waldorf salad when hit with the pressure.

Finally a proper fit was achieved and an orange launched out of the tube with a 400-psi fffwhump. It soared into the blue sky like a home run out of Candlestick Park, disappearing over the top of *Vinson's* flight deck, 20 stories up. We submariners cheered and yelled insults at the carrier.

"Take that, skimmer pukes!"

"Motherfucking Target!"

"Don't make us come over there…!"

The celebration quieted then died, as heads appeared over the edge of the big ship. First one head, then three, then forty. Some heads were wearing hats with the look of authority. All were looking in our direction.

A bit of aircraft carrier trivia popped into my head. Aircraft carriers do not take kindly to foreign objects on their flight deck. Foreign objects do bad things when sucked into aircraft engines.

It was an excellent time to be elsewhere. Keith said "fuck" and in two heartbeats the fierce topside warriors vanished. We dropped down the Engine Room and Stern Room hatches, sliding down the ladder railings without touching the rungs. We busied ourselves below deck for anxious minutes until lines were cast off and *Seawolf* was away from the pier.

We had met the enemy and he was ours.

22

Race Relations

Penetration, however slight, is enough to complete the act.

— *Uniform Code of Military Justice, Article 120*

A TYPICAL SUBMARINE crew was comprised of two-thirds twenty-somethings, one-third thirty-somethings, and only a handful of men over forty years of age. Only a small fraction was younger than twenty years because most submarine billets — Navy talk for job positions — required extensive schooling. Even though they enlisted at eighteen, they were twenty-year-olds by the time they were assigned to a submarine.

The oldest men on any boat were likely to be the Chief of the Boat and the Captain. On *Seawolf*, Chief Warrant Officer Wiseman, CWO4, had started Navy life enlisted and moved up into officer-hood. Mr. Wiseman was a grand gentleman that reminded everyone of a beloved grandfather. Gray-headed and soft-spoken, he led quietly by example. There was something about Mr. Wiseman that made you want to not disappoint him. He was the oldest man on the boat. He was probably in his late forties.

Nukes tended to shift the average age on the boat upward. Many had two or three years of college under their belts before enlisting. Having run out of out of money or motivation, they had sought alternative ways to fund their education or do something meaningful. Many had families and had held jobs out in the world prior to enlisting, but not all. Plenty of Nukes signed up straight out of high school.

A young sailor could find himself working with—or for—a man with a heritage that might spawn racist comments in his hometown barbershop. Racial bigotry, by policy, was not tolerated in the Navy. However, official decree has never changed a mind, only public behavior. It takes time and experience to undo the conditioning of even a young life.

I never observed true racism in the submarine navy. In using the term I refer to a loss of opportunity or status due to a person's race. I am not including coarse talk which, while disallowed and reprehensible, did not limit the maligned man's paycheck or advancement.

Men were judged on their knowledge and character, with character trumping all else. If you want to be treated as a first-class guy on a submarine, you learned your business and didn't shirk a task. No one cared who your father was or what neighborhood you grew up in. No one cared what color your skin was. On a submarine, you defined who you were.

Stand-up guys come in all shades; black, white, pinstriped, polka-dotted, and everything in between. So do jackasses.

Machinist Mate First Class John Parrott, a man of Hispanic ancestry, liked to whack political correctness straight on the nose by confiding to junior officers: "Sir, you're my favorite white officer."

The young officer in the spotlight wasn't sure if he was being complimented or insulted. The guy was respectful and seemed appreciative, but…? What to do, what to do. The LTJG usually studied his shoes until he remembered where he was needed and hurried off.

Parrott would smile and holster his little conundrum. It worked every time to confuse and stupefy junior officers. Trying the same trick on wiser men usually resulted in a response to the effect of "go fuck yourself." That too, was a response to which Parrott could laugh.

Since Navy men hail from every part of the country, every sailor has experienced trouble with a language barrier. Often it involved Filipinos. The Navy carried a solid contingent of Filipinos, I suppose naturally occurring due to our WWII involvement in the Philippines. A heavy Filipino accent becomes more pronounced with anger. The Filipino speaker becomes increasingly enraged by the non-Filipino sailor's inability to distinguish what the speaker is saying.

In San Diego—and this is no shit—I heard a Filipino chief petty officer, with an accent so thick that words were indiscernible, scream at a confused seaman, "Don't you understand English?"

The seaman calmly replied, "Yes, I understand English. Can you find someone who can speak it to me?" That was the end of the polite part of that conversation.

Once a U.S. Navy ship is commissioned it is continuously manned by enough qualified people to man every watchstation necessary to put the ship to sea. The nuclear reactor is attended 24 hours a day for the life of the ship, even when shut down, even in overhaul when the nuclear fuel has been removed and the reactor vessel sits empty.

In port, most of the crew went home after a normal workday. A Duty Section stayed behind, minding all the things that needed minding through the night. In four-section duty, I went home after work like a regular civilian for three nights; on the fourth I had "the duty." I worked a regular work-day, stayed the night, working much of it, and worked the next full day as well. A duty day was a thirty-three hour day, more or less, with perhaps a few hours of constantly interrupted sleep thrown in as a perk.

Need I mention that Navy personnel haven't a union?

On a Tuesday evening, *Seawolf* was in port and I had duty. Electrician's Mate First Class Clarence Davis and I were sitting in the Crew's Mess. His friends called him Clem; I never knew the origin of the nickname. I've known Clem 27 years at time of this writing and we have shared a few adventures together post Navy. I still don't know how the moniker of Clem came to be. Clem's a private guy who's masterful at maintaining that privacy. Professional spooks could learn a thing or two from Clem. Who am I to screw with his secrets? How he got the nickname is probably a lame story anyway. (Ping, Clem.)

The Crew's Mess was a rectangular space twelve feet wide by twenty feet long. Five stainless steel tables were bolted to the deck; blue Naugahyde cushions topped their bench seats which accommodated six to a table. (If you've forgotten *Seawolf's* age, I have seen a photo of President Eisenhower eating a meal seated at one of these very tables.)

A group of men were having a late supper. They laughed and joked and dug into their food with gusto. They were all around twenty years old.

Clarence and I were seasoned salts at the advanced age of twenty-six, well on our way to being "Old Guys."

In the course of their horseplay and banter, one of the young men commanded another to pass the potatoes. The potato-passer snarled back, "Who was your nigger yesterday?"

Startled by the excruciating realization that they were not alone, the table mates collectively hunched their shoulders as if awaiting a blow.

You see, Clarence Davis is a black man.

Six hang-dog heads creaked around to look toward Clem, but they couldn't look him in the eye. Their white faces were blanched whiter with the shame of having been caught in so juvenile a display.

It struck me that they were more worried about having hurt Clem's feelings than about any repercussions they might suffer. Mortified as they were, if the Captain had ordered them shot out of a torpedo tube for their sins, I believe they would have agreed to the verdict.

I respected them for their suffering. If not better discretion, they at least had working consciences.

Clarence Davis is a stand-up guy. No better man could you want on your submarine. He also knew there was no correlation between the man that he was and a rude word used by silly people.

Still—the words hung in the air like a mushroom cloud: Who was your nigger yesterday?

Clarence Davis looked at the men impassively over his cup of steaming Earl Grey, letting their misery chafe and burn for several tense moments. Then he flashed a brilliant Clem smile.

"Well," he said, "it wasn't me."

Most people don't know what their "hundred-percent" looks like. Many times in their lives everyone has sworn to do their best, to give one hundred percent to some endeavor or another. Every member of a sports team has joined hands and vowed to give their all on the field. Possibly they did give all. But how do they know if they had actually given all they had to give?

Most of us have been able to excuse away our shortcomings. Miss an assignment deadline and all that is required is a little begging and you'll get a new deadline. Reluctantly, perhaps with a penalty, certainly with a stern warning, but you'll get a second chance. A special dispensation will be granted; just for you, just this once. It's not your fault, it's the system.

American society is bent on dodging accountability. We accept that fast-food corporations made the kid fat and tobacco companies forced cancer on unsuspecting innocents. Willpower is a made-up word to help ladies be charming in proximity of Nordstrom or chocolate. We are used to stopping as soon as it hurts, resting when tired, and eating when we aren't hungry. Many people working out at the gym never break a sweat. They're frustrated because they've been going to the gym regularly and haven't lost weight. Mankind shares a self-limiting habit of asking everything of others and very little of ourselves.

"This is your first day of school, Johnny. Just do your best, that's all we'll ever ask of you." That may be fine encouragement for a five-year-old, but somewhere along his way to manhood Johnny needs to find out what he can do when he puts his back into it. What is his best?

When I see a young man after Basic Training, I am usually pleasantly surprised. The boy I knew is no longer. That boy has been replaced by a man. Regardless of which service he joined, he stands tall. He speaks respectfully and with confidence. There is purpose in his stride. He has neither been brainwashed into submission nor into blind obedience. He still doesn't know where the limit of his hundred-percent is—very few of us ever do—but he knows for damn sure that his limit is way higher than he thought. His new confidence is a direct result of events like one I witnessed on the running track, back in Basic Training.

My Recruit Training Company was to perform a physical fitness test. This run was one of the last major hurdles between us and graduation. We had begun to feel hope that we might survive boot camp and were getting cockier by the day.

Before basic training, sitting at home on the sofa, it had been easy to calm my anxieties about going to Boot Camp. It would be tough I knew, but they wouldn't kill me. They couldn't go around killing recruits. By the second week of Boot Camp that comforting assumption had vanished. Now in the final week of Basic Training we thought we were tough for surviving this long. We were ready to kick that two-mile run right through the goalpost. Bring it on.

A Senior Chief Petty Officer stepped out of his office at the athletic field to inquire why this recruit company was standing around with the unmistakable look of sheep without a shepherd.

We told him we were there to perform the two-mile run for time. No, we did not know the whereabouts of our instructor.

We milled about uncomfortably, a rookie mob of no authority debating the proper next move. Mostly we were scared. In Boot Camp we were constantly under the direct control of someone who knew exactly where to be and what to do. Now we were alone. Something was clearly amiss. We assumed we were in the wrong—a normally safe assumption for recruits—and debated whether we would suffer more for being at the wrong place at the wrong time or for not speaking up.

The Senior Chief spent thirty seconds ascertaining the situation and trying to find out which instructor was supposed to be there.

There was confusion on that issue. Someone on his staff suggested that one of them should watch us run and verify our satisfactory time.

Without hesitation the Senior Chief said, "I'll do it."

It is significant to note that the Senior Chief was dressed in crisply starched winter working blues, complete with necktie, service ribbons, and brilliantly shined dress shoes. His long-sleeved black uniform was tailored to perfection. A white combination cap with a black patent leather brim set off the uniform brilliantly in the San Diego sunshine.

Not a tall man, he was trim and looked solid. A few murmurs in the ranks of "pint-sized" and "short stuff" were hushed with a low hiss from the Recruit Chief Petty Officer, who was one of us. The RCPO was designated as our leader when the company commander was not present. We didn't respect him any more than the Easter Bunny but we feared the suffering our company commander would surely deal out if we bucked the RCPO.

Many individuals can run two miles in twelve minutes without much trouble, especially young men eighteen to twenty-two. However, eighty men running in formation, in step, is a different situation. It's particularly tough on the long stride guys and on the short stride guys. If you are five-feet-six-inches tall you have it made, stride-wise. As a long-legged guy, I would rather run ten miles on my own than run two miles in formation.

We would pass or fail as a company. No one could lag or fall out to suck wind. The jocks, the marching band, the chess team, and the debate team, all must cross the finish line together. One misstep could trip up and lay out forty men. If someone passed out from heat exhaustion, they would be carried. If someone needed to vomit, they would do it in the ranks without slowing. That would make those within the splash zone unhappy, but fall behind they would not.

The squared-away Senior Chief ran backward in front of us the entire two miles, railing at us for our careful pace which he thought too slow.

Crossing the finish line only a few seconds ahead of the time allotted, we fell on the grass, our chests heaving. The Senior Chief stood straight, hands on his hips, crisp and sharp. Not panting like the rest of us, he spoke while we recovered.

Quieter now, man to man, he encouraged us to become outstanding sailors in his fine Navy. He admonished us to never accept "good enough" but to set a higher standard for ourselves and to require the same of those around us. He had our full attention and we listened.

The San Diego sun was hot on his long-sleeved uniform, yet he ignored the sweat on his forehead. Navy uniform regulations do not authorize a loosened necktie or an unbuttoned dress shirt collar. Therefore, the Senior Chief did not loosen his necktie. Even when running in the sun.

It may sound trite, this story of a man who ran an athletic test in attire inappropriate for the event. Perhaps I'm not relating it well; perhaps it's one of those times you just had to be there. It would be easy to think of that Senior Chief as ridiculous. Loosen your tie, man, and strip down to your t-shirt, put on some sneakers. Who would care? Or he could have taken a few minutes and changed into running gear.

In the years since, I have tried to nail down why this brief episode affects me so, why it remains so vivid. This memory is not about athletic prowess or clothing. It's about character.

The man stood up to the situation confronting him. He did not hesitate to take charge when initiative was needed. He quickly gathered information, he made a decision. He acted and lived with the discomfitures of aggressive pursuit of the desired outcome. He did not retreat toward comfort or convenience. He did not shirk nor snivel.

All philosophizing aside, it could also be that I just liked the guy. He was a stud.

Personal excellence has been demonstrated to me many times in the years since, but I remember that Senior Chief running backward in a dress uniform, blasting our "minimum effort" in his "four-point-oh" Navy.

Eighty percent of Recruit Training Company 946 was white. That Senior Chief was a black man. There were zero-point-zero thoughts of bigotry among us during and after that run. How you do one thing is how you do many things. There was nothing minimum about the man. Whatever color he was, who his father was, what neighborhood he had been raised in, was inconsequential. He defined himself.

I never saw that Senior Chief again. I know that I am a better man for having spent twenty minutes of my life with him.

23

Class Warfare

[Pertaining to boats not submarines]
Enlisted men who are passengers in running boats which contain officers shall maintain silence.

— *Bluejackets' Manual, 1940 edition*

ABOUT A YEAR ago I attended a 4th of July party set in a park in one of the nicer neighborhoods of Portland. Seated at a picnic table by the pond with several people the conversation turned to what had transpired to bring each person to Portland. The husband-half of an older couple told of being stationed in the area during WWII. He had been an intelligence officer.

Many men slip from reality into fantasy about past military service. Any bar will have at least one guy who talks tough about Special Forces and the 'Nam. Later you will likely find out that he was an Army auto mechanic in Germany for his entire time in the war.

Not long ago I worked with a man who told several others in the company that he had been a Navy Seal. At a company Christmas party his wife told me he had been a Damage Control technician. This is the Navy version of a firefighter. Being a Damage Controlman was a fine profession, but he was not what he advertised. Apparently he hadn't briefed his wife on which past he preferred, the real or the imaginary.

But I believed that my table mate at the 4th of July party had worked in intelligence as stated. Someone else at the table asked him what projects he had worked on during the war and his answers and manner became vague in a way with which I was familiar. It was a peculiar vagueness, not the glib answers of someone without a clue or of someone making it all up. He had that friendly look that said, "I like you well enough, but you and I aren't going to talk about that." His was the precise vagueness of a man compelled to secrecy.

Another man at the table had been an Army officer in Vietnam. The two men got along wonderfully. Finally, the intelligence officer's wife turned to me.

"Were you in the Army too, Karl?"

No, I told her. I had been Navy.

"The Navy! How wonderful!" she responded. "What did you do in the Navy?" I told her I had been in submarines.

"Submarines! How interesting!" she gushed. "How could you do it, being underwater all that time?" I told her it was like taking a long trip in her car. You get in and shut the door. You drive. You don't get out again until you stop. She thought this banal explanation was terrifically clever. I had used it before. It usually satisfied the listener.

The intelligence officer's wife laughed. "So smart! You must have been an officer!" she said admiringly.

No, I had been enlisted.

"Oh, I see." Her tone became the same tone my father had used when I gave him a necktie with a trout painted on it for his birthday. My father never wore neckties. He was a once-a-year fisherman. And the painted trout, well...no one should have worn that tie.

Sometime in your past you've heard the tone that was in his voice on that day. "Thank you... what a great gift...I'll cherish it..." Next.

The intelligence officer's wife withdrew from me and took up conversation with across the table. I imagined her reaction would have been the same if I had said that I was a convicted felon on parole. I decided to use that answer the next time I was queried at a party.

I had experienced this reaction before today. When talking with former enlisted people of any service branch, they ask about what was my rate and rating—about my job in the Navy. Where did you go, what did you do? When talking with currently serving officers, former officers, and officers' wives, the spirited conversation usually ended at "enlisted."

I doubt the offenders, or perhaps rather the offended, of this lower-tier social standing of being enlisted meant any insult. It simply surprised them. I think they were stunned by the idea that they had been having an intelligent, relevant conversation with a military member who was not an officer.

Clearly they hadn't known some of the officers with whom I've tried to have relevant conversations.

Hollywood's standard depiction of enlisted sailors is cartoonish. A typical movie plot runs like this: The Commanding Officer, a handsome leading man, is brilliant yet suitably conflicted. (We will skip the standard unrequited love interest.)

The crew, barely capable of independent thought, looks to the Captain in every crisis.

"Golly, sir, what should we do now?"

Even films determined to treat the workings of a warship seriously do not portray the crew accurately.

All organizations, civilian and military, make distinctions between front line and senior management. The front line eats in the cafeteria while the big guys in the executive dining room eat a fine meal provided by the executive chef. The salesman fly coach; the C-suite are flown in the corporate jet. The salesmen rent a car; the executives have a car service waiting. The workers make a low multiple of minimum wage and management collects six-figures plus bonuses for their excellent managing of your hard work.

We know these things are partly a matter of choices in life and partly a matter of who your father is. If he has the sway get you into the Naval Academy, you are on your way. If he doesn't, and can't send you to college, you'll have to work your way through or begin your career sweeping the floor. This is the way of the world.

Nowhere have I experienced the difference between blue collar and white collar more obviously than in the U.S. Navy. This divide is highlighted by the close proximity of the haves and the have-nots on a submarine.

Officers slept in staterooms, even though most officers were two to a stateroom. It was not spacious, not even what one would expect on an economy-class cruise, but it was a semi-private room with a door. Enlisted men slept in a bunk room with dozens of others. On some submarines, they shared beds. One man worked while another man slept.

An officer enjoyed meals served to him on china plates. It wasn't the finest china, but it had gilded edges. The officer ate his dinner using silver utensils. He wiped his chin with a linen napkin. An engraved silver napkin ring marked his place at the table. Enlisted ate with stainless steel picnic-ware on plastic plates after standing in line for a seat.

Be advised: if you enlist after you have experienced some life, had a job, and paid your own way, you will be required to subordinate yourself to a 22-year-old whose father held enough blue chips to get his son into Annapolis. The young officer cannot yet grow a mustache, he may or may not be particularly smart—someone is last in every class—but he is your master. No one is more aware of that fact than he.

A military salute, as explained to me once by a Navy captain, is a mutual sign of respect and comradeship. He explained about medieval knights lifting their visors for recognition and as a sign of peaceful intent. It was an honor to render a salute to a comrade in arms. I felt good about that until I thought on it further.

If a salute was mutual, why was I required to salute the officer first? He merely returned my salute. Saluting an officer seemed to carry a similar significance as bowing and scraping. I was all for respect, especially the mythical mutual respect, but genuflecting seemed over the top.

This act of respect and submission is carried on through the officer corps. An officer of junior rank must salute a senior in the same manner as enlisted people must salute any officer. The uniform differences between officer and enlisted could be easily seen at a distance. The uniform differences between officer ranks were more subtle.

Why must I say "Sir" and "Ma'am" to officers, while in the name of mutual respect they call me "Petty Officer?" Let's not even start on the term "petty."

If I don't salute properly, I stand to be punished. Punished! Anything that one can be punished for not doing shoots down the concept of mutual.

While in school at Naval Training Center San Diego, I was paying too much attention to the base map in my hands and walked past an officer without saluting. He stopped me and delivered a thorough and harsh dressing down. He was red-in-the-face mad. He wrote down my name and the school I was attending so that he could ensure disciplinary action was taken against me. I think he may have been a commander. I don't recall the man, only the incident.

A sign of mutual respect and honor. Hmmm…

I didn't see the sense in the man going off like he did. To correct me, to instruct me, sure, I would have had no problem with that. But he was angry that I had somehow cheated him personally and he demanded satisfaction. He had demanded my humiliation; he put me in my place. Had it been two hundred years earlier I have little doubt that he would have had me whipped on the spot. He would not have challenged me to a duel over the slight, as I was clearly not of his class.

Grow up and get over yourself, we've both got better things to do, I told him. Only in my head, I said that only in my head—not out loud, hell no—only in my head. In fact, it was several hours later that I thought up that snappy retort, after I had stopped shaking with rage.

That small incident stayed with me. That officer's petulant tantrum had as much to do with me eventually leaving the Navy as would any other single issue. I didn't think grown men should behave that way regardless of incitement. I believe that still.

When listing a ship's compliment, or on a monument to a lost ship, the Navy always listed separately Officers and Men. The enlisted me joked that this was proof that officers weren't men.

That traditional phrase always bothered me. It implied a separation larger than earned rank and job description. It represented the opinion of the U.S. Navy that I, an enlisted man, was so much less than an officer, that should I die a horrible death in the service of my country, my name would not be intermingled with that of commissioned officers. The officers' names would come first, then we others. After all, what would the neighbors on Officers Row think, if one of a lower class should be listed with their people?

Would such a travesty cause outrage and uproar from the Officer Corps? Would they petition a politician to change the plaque? Maybe an admiral would see that an asterisk was placed by my name with the footnote: "Not one of us."

Seawolf was in one of her frequent periods in the shipyard. Scaffolding ringed the weather deck topside. Lieutenant Junior Grade Charles (not his real name) was the Engineering Duty Officer. (The irony of my diatribe on the officer/enlisted relationship while this poor guy's title contains the term "junior grade" is not lost on me.)

When greeted with a friendly 'How's it going,' LTJG Charles let out a ponderous sigh, and with great weariness reply, 'Oh…it's going okay, I guess.' "Okay" was as good as it got for this JG. We called him Eyore. Not to his face, of course.

Late on a Monday evening Eyore climbed wearily down the Stern Room ladder and shuffled down the passageway into the Engine Room. The Lieutenant informed the Bull Nuke of a Coke can topside near the sail. The Lieutenant instructed the Senior Enlisted Engineering Department Representative that he should inform the Aft Duty Section Leader of the atrocity and have the offensive can removed. The Bull Nuke, nicknamed Snaggletooth for a wayward front tooth, informed the Lieutenant Junior Grade that the aft duty section leader was in Maneuvering. If the LTJG was going in that direction, he might tell the duty section leader his own goddamn self, especially since the LTJG had walked past Maneuvering to find the Bull Nuke.

Chagrined at the lack of esteem he was beheld in, but knowing only how to suffer, Eyore shuffled into Maneuvering to locate Petty Officer Timmes (not his real name). In addition to being the Aft Duty Section Leader, Petty Officer Timmes was on watch as the Shutdown Reactor Operator. Slumped in his seat in front of the Reactor Plant Control Panel, the SRO slowly turned his head to acknowledge the EDO.

"Petty Officer Timmes, are you the Aft Section Leader?"

"Yes, Lieutenant, I am."

"There is a Coke can topside near the sail. Have someone from the duty section remove it to the rubbish bin."

The Aft Section Leader coolly contemplated the Engineering Duty Officer.

"Remove it to the rubbish bin?" the Aft Section Leader asked incredulously. Choosing to mock the man in another way, he let go his initial reaction to the affected language. He would save that abuse for later. "Is the soda can empty?"

"I don't know," replied the LTJG. "That doesn't really matter does it?"

"Well sir, if it's empty you could save it. You could put it in your pocket and take it to the grocery store. You could get a nickel for it. You could send the nickel to your father to repay your college tuition at whatever prima donna Ivy League university he sent you to," the Aft Duty Section Leader responded with a wide-eyed innocence devoid of insubordination. Or should I say, he responded with shrouded insubordination that would be difficult to prove in a civilian court where evidence carries weight.

As an afterthought the SRO added a stinging accusation as a fisherman deftly lays a dry fly into a dark riffle on the barest chance that a fish will rise.

"You were in the Glee Club weren't you?"

Eyore glared at the Aft Duty Section Leader.

"You *were* in the Glee Club. Damn, sir. I used to think highly of you." The SRO continued feigning befuddled innocence and pressed his attack.

"Besides sir, I'm kind of busy right now. I am the SRO on watch, as you can see. You can tell because I'm the one sitting in front of the RPCP at a quarter past midnight, while the normal people out in the world are sleeping in their warm comfy beds."

LTJG Eyore was annoyed. "I can see that Petty Officer Timmes. I meant for you as Aft Duty Section Leader, to instruct someone else in the duty section to do it."

"Sir, you want me to wake somebody up, at a quarter past midnight, so they can walk 80 feet to pick up a Coke can tossed by some dickless yardbird that you could not soil your lily-white fingers on? You were just there, why on earth didn't you just pick the damn thing up and throw it away? You passed two trash cans on the way here!"

The Lieutenant tugged on his collar calling attention to his single silver bar. The SRO lifted himself out of his seat to get a closer look. He studied the LTJG's collar device and let out a low whistle. He smoothed the chevrons on his own sleeve with his middle finger.

"I guess you got me there, sir. One silver bar beats three chevrons all day long."

Eyore looked satisfied. "What time it is doesn't matter. The Navy does not operate on the nine-to-five," he said smugly. But the weight of LTJG's responsibilities returned quickly. He sighed deeply.

"Being in command doesn't make me popular," he said.

"In command?" the Aft Duty Section Leader/Shutdown Reactor Operator raised his eyebrows. A fresh new lieutenant junior grade was about as far from command on this nuclear submarine as was Mickey Mouse. This man had set a record for taking the longest to qualify EDO. This midwatch was becoming downright entertaining.

"I meant as in being the Engineering Duty Officer. I am in command of the Engineering department. At the moment." The SRO slowly swiveled around to face his panel. He was too tired to joust with Eyore. This witless banter was wearisome; he was engaging in a battle of wits with an unarmed man. He was tired, but he would have risen to the occasion for a more formidable adversary. The EDO left Maneuvering to make his routine tour of the engineering spaces.

Now it was the SRO's turn to sigh deeply. This LTJG could be a commanding officer one day. The world was going straight to hell on a runaway elevator.

Relieved of the watch at 0530, the SRO climbed up through the stern room escape trunk and out the hatch. He strolled to the sail and found the empty soda can. Instead of putting the empty soda can in the trash can—the fucking rubbish bin, he thought—he crushed it flat. He flung the aluminum disc as far as he could out into the river.

He drew a deep breath of the early morning air and leaned on the rail of the temporary scaffolding that lined the submarine topside. He pondered the lights of Vallejo across the river. The horizon was crowned with gray, but the town was still dark. There were lights on in some of the homes on the hillside facing him. He thought of the still warm beds recently vacated, their covers thrown back. He imagined women in bathrobes shuffling into kitchens to make coffee, and pack lunches for the kids. He could almost smell the coffee brewing.

He thought about his own wife. She would be getting up in a few minutes. She would wake the children and prod them along in their morning rituals. Hurry now, you'll be late for school. Honey, I don't know where your shoe is, why isn't it with the other one?

He found himself smiling at the thought of his kids. They woke slowly and would be eating their toast half asleep, their legs swinging from the chairs around the kitchen table. His daughter liked to prop up her sleepy head in one hand, the elbow resting on the table while she chewed with her eyes closed. His wife would stroke the child's tousled hair away from her mouth. His son would whine a little, about what the father was never quite sure. The boy just needed to fuss a little in the morning. The father suddenly realized that he knew just how the boy felt.

He chuckled out loud at this last thought. He had two years left on his hitch in the Navy and sometimes he felt like it would never end. Being on the submarine had been tough on his marriage. His wife and he had enlisted as a team, he as a Navy man and she as a Navy wife. But the hours were so brutal. They challenged everyone's resolve. He was 72 hours at work on the light weeks when the boat was in port. When he had two duty days in the same week he was at work 96 hours.

His wife didn't understand why he wanted to sleep on his few days off. She was frustrated by having to keep together house and home, and raise their two kids, seemingly by herself. Her unhappiness made him unhappy.

Sometimes he just wanted to whimper like his little boy did in the mornings. But he was a grown man. He knew all the tantrums in the world wouldn't change a thing. He would survive his 8 years in the Navy. He wondered if his marriage would.

After a few minutes of reverie, he inhaled a double lungful of salt air. He held it for a moment, and then let himself deflate completely. Everything would be alright.

He regretted throwing the Coke can into the river. Just what the world needs, he thought. One more asshole trashing up the place.

He climbed back down into the boat to begin the day. He was already exhausted. His day had really started six hours ago, when he took the midwatch the citizens of Vallejo slept, but that didn't count. The Navy does not operate on the nine-to-five.

The first time I thought I might die on this submarine I felt wonder and sadness. We were at sea and had suffered a major casualty. Obviously we fixed it and came back safely, but I was changed. I had experienced a moment like I had not experienced before.

At the moment when I realized that the danger was high enough and imminent enough to do us in, I remember feeling disappointed. There were people I loved and wanted to see again, and things I wanted to do. I was swept over with wonder that this was how it would end, like a fly in the kitchen. Flying free one moment; swatted out of existence the next.

On the second occasion I thought we might not return home ever again, I was angry. Angry that my final hour was going to be spent with these assholes on some mission decreed by assholes in Washington D.C.. Angry that my daughters would be fatherless.

In either case I don't know if our situation was truly dire or if my mind was replaying a "Star Trek" episode. The sailors on the starship Enterprise were about to be sucked into the abyss once a week. Nevertheless, the thought had flashed through me that the next few minutes could possibly be all we would get.

No nation with an IQ above 80 at its helm would take on the U.S. in an all-out, face-to-face war. That time is past. The way to take down a country in the 21st century is economic and electronic, a stream of 1s and 0s wreaking havoc. The tools of destruction in the most modern of wars will be banks and interest rates. Poison their economy, buy their debt, and then serve an eviction notice on the seat of power. No credit, no resources; they will capitulate, or they will wither and die.

The power brokers sit in upholstered swivel chairs and use Armies and Navies quietly, as pieces on a global chessboard. When "quietly" is no longer possible, militaries become politically expedient. When nothing else works, send in the 7th Fleet. As a plumber uses the correct wrench to stop the leak, that image of power and strength may be the perfect tool for the job.

I'm sure the politicians agonize over the decision to use military force. They certainly agonize over the media coverage.

They make the decision. The Commander in Chief wills it.

Then the politicos order a fresh Martini, with Extra Dry, please, and excitedly plan their eight-day stopover in Milan on the way to a six-hour fact-finding tour in Afghanistan.

If our submarine had been sunk in an open act of aggression in response to our covert acts of aggression, the whole event would last one news cycle.

I imagined the news of our recent demise would arrive during a business lunch. At the Old Ebbitt Grill, a short walk from our nation's capitol, news of our tragedy would be whispered into the ear of the Senior Senator from California. California is *Seawolf's* home state; it would be natural to inform her. Many of the lost would be California residents and registered voters, every one.

She would whisper instructions for her staff. Prepare a public statement, full of regret for the loss and gratitude for our dedicated service. The statement should assure the country that the culprits will be punished. Put lots of tough talk in, she would instruct. And find a way to place blame for our deaths on the other political party. She asks; what was the name of that ship again? It was a submarine?

She then would apologize to her lunch companions for the intrusion. She would return to her Shrimp Casino and Napa Valley chardonnay. Her lunch tab would, of course, be picked up by the lobbyist wooing her for support. This is not a game for cheapskates. He is probably promoting something detrimental to mankind in the long run, but good for profits in near future. His industry will donate the appropriate amount of money to her reelection fund. His company also has a luxury condo in the Bahamas, complete with staff, that she and her family are free to use as often as they wish, with the compliments of his board of directors.

The bill will pass. The loss of the submarine with its 125 men will pass through the news media as fast as a bottle of Fleet. The loss of the ship will be played off as an accident. The political ramifications of revealing the mission are too damaging.

And so it goes, pawns are lost but the agile back row remains on the board.

24

Deck Gang

No battle plan survives first contact with the enemy.

— *Military axiom*

Seawolf PULLED INTO Mare Island after several weeks at sea. When the ship came into or left port, the Maneuvering Watch was stationed. The idea of the Maneuvering watch bill was to put the most senior watchstanders at their posts during the critical period of maneuvering the ship in tight quarters. As open water around the moving submarine diminished, the chance of inadvertently running into something undesirable increased proportionately. A slow-moving submarine on the surface was a four-thousand ton blunt instrument looking for something to smash into.

Normally I would have been stationed on the engineering department maneuvering watch bill. On this occasion I found myself listed on the ships maneuvering watch bill as Number Three Line Handling Captain.

This was a pleasant surprise. Being at sea meant no sunshine. This assignment would put me up on the weather deck as we came in. Outside! Wow!

There was a hitch in this gig. I knew nothing about deck seamanship. Ropes and knots are the purview of boatswain mates and surface ships. Skimmers, Targets, whatever you called them, they still used that old marlinspike seamanship stuff. I was a Nuke; I dealt in neutrons, not rope. The closest I ever got to a knot was tying my shoes. Lines were Coner work, Forward Puke stuff, not Nuke stuff. Nevertheless, I was now a line handling captain.

Pleased that my leadership qualities were finally recognized, I puffed a little with the title. Number Three Line Handling Captain. I had never handled a line in my life. Was this part of the vast encyclopedia of stuff I was supposed to know but didn't?

The order was passed; line handlers lay topside.

After being submerged for weeks, everyone topside was in the best of spirits. This was living large.

I found the Forward Line Supervisor, a Chief Torpedoman.

"Thanks for giving me this job Chief," I said. "I'm curious. How did you choose me?"

"I am looking to put an infusion of quality into Deck Gang. I know you are just the man to do that."

Well, goddamn. I swelled a little more.

Finding line Number Three required some sleuthing work. I had never looked upon mooring lines as anything I would ever handle. My knowledge of them was enough to pass the written exam, but I failed the lab.

My line handling crew was waiting. Number Three mooring line was out of its locker and faked down on the deck. Faking a line is a way of laying it out so that it will run freely without snagging when it is thrown. I can say that now. At the time I only knew the heavy mooring line looked cool laid out in flat coils on the deck. I thought it was showmanship. It wasn't. My line crew got snippy when I thoughtlessly stepped on the coil.

"Get off the line," they snarled in unison. Disgust was clear on their faces.

Not wanting to reveal my stupidity to senior people supervising the mooring—it's possible that ship had already sailed—I chatted with my line crew like an old hand. From them I discovered that assignment for line crews was determined alphabetically. I looked at the line handling captains forward of my position.

Line One captain was a guy named Adams. Line Two—Downing. Line Three—Heckman.

I looked aft. Line Four captain—Johns. Line Five—Morrison. Line Six—Zemetski.

Infusion of quality my ass. I was going to keep an eye on the Chief Torpedoman. A guy who could sell bullshit that easy was dangerous.

My line crewmen were all forward types, non-nuclear trained personnel. Prejudicial, yes, but I assumed that as forward guys this line handling thing was their bailiwick. Surely they had studied this, passed checkouts on it and knew what to do. I came clean to my line crew.

"I've never done this," I confided. "I don't know what to do here." They nodded.

"No shit?" one said. All were grinning. They shook their heads in mock amazement. "Fucking Nukes..." The first speaker spoke again.

"We got this. Just try not to get hurt, you useless fucking Nuke."

"Cool," I said. "Thanks. Please don't let me fuck up." They were grinning again.

"Shaka brah, we gotcha covered," they said.

Leaders are born and leaders are made. Some are determined alphabetically.

My anxiety lessened, I waited for orders and reveled in the breeze and the sunshine.

The comings and goings of a nuclear submarine are secret. Wives don't know when husbands will return. It may be weeks, it may be months. When the boat was heading in, a day or two out, a "call tree" was activated to inform wives that their men were coming home. The Ombudsman, a wife of one of the men on the boat, would call two people. They in turn would call the two people they were assigned, and word was passed.

When the boat rounded the southern end of Mare Island and passed the Coast Guard station and past the slips where DevGroup submarines *Russell* and *Parche* would have been moored had they been in port. A crowd of women and children waited on the pier near berth 19, our Mare Island reserved parking spot. In their Sunday best the wives and kids were a welcome sight.

Still too far away for shouting, sailors and families waved at each other.

"Jones, isn't that your wife? There next to the light pole in the yellow dress."

"Hey, there she is! Hi baby!" Jones shouted into the wind. His wife enthusiastically waved back. People who didn't even know Mr. and Mrs. Jones smiled. Happiness is contagious.

"Dave, there's your girlfriend. She looks great. If you have duty tonight, I'll take her out for you."

"Thanks a lot, asshole. You're a real pal. Get away from me."

Going to sea was a big pain for a family, but the homecoming was almost worth the inconvenience.

We smiled and laughed in fanatical joy at the little girls in frilly dresses that fluttered in the wind. Boys jumped up and down to draw their Dads' attention.

A new baby was held aloft, giving it and the new father a first glimpse of each other. Seeing his son for the first time, the young sailor waved and smiled wide. A shipmate clapped him on the shoulder and pointed to the baby.

"That's your son, Charlie! Your son!" The young father's emotions broke free. He clapped his hands to his head as tears streamed down his face. His buddies surrounded him. They didn't have to say anything. An arm went around his shoulders and held him up until he steadied himself. Everyone applauded.

The scene made me misty. Just a little, mind you.

"Damn wind in my eyes," I complained to the Chief standing next to me. He was watching the new father and smiling quietly, like a man who knew.

Without looking at me, he said "Yeah, that wind will get to you every time."

A call rang out:

LINE CREWS, STAND BY YOUR LINES

Aye, aye. Standing by.

A tugboat sidled up to the boat and was made fast. The tug pushed *Seawolf* sideways the last few feet to the pier.

Line handlers on the boat warmed up to throw heaving lines to their counterparts on the pier. A heaving line was made of light rope and with a complicated knot at the end called a monkey fist. The knot was the size of an apple and added weight to the end of the line. With practice a heaving line can be hurled a long way.

The Bluejackets' Manual of 1940 encourages a sailor to practice throwing a heaving line "at every opportunity." In the 1980s Navy I never noticed having the opportunity until now and this wasn't practice but the real deal.

Men grasped in one hand the heaving line about two feet from its end the monkey fist dangling. Their other hand held several loose coils of heaving line with the balance of the light line attached to the heavy mooring lines flaked out on the deck. Almost simultaneously but not quite, resembling an Olympic synchronized line handling team that didn't make the cut, heavie throwers flipped their wrists and swung their monkey fist in a circle like a cowboy swinging a lasso vertically. When satisfied with their guesstimate of the line's likely trajectory, they let fly. Some throws were more accurate than others. I heard "look out!" yelled as an errant heavie arced through the blue sky toward the crowd of onlookers.

Line-handlers pier-side caught the heavies like an outfielder brings in a fly ball. The monkey fist would raise a knot if it coincided with a human head, a couple of the wiser pier side line handlers let the thing land unimpeded on the concrete and then stepped on the line preventing it from sliding off into the water.

Pier side men pulled in the heaving lines followed by the heavier mooring lines. Pulled across to the pier, the looped end of the mooring line was dropped over a bollard, an iron post mounted on the edge of the dock. On command, the line handlers on the boat pulled the lines tight. The submarine was snugged up to the side of the wharf.

Once the submarine was positioned the lines were doubled up, then tripled. Each mooring line crossed from pier to boat, back to the pier, then back to the boat and made fast around a cleat built into the deck. Six lines tripled. *Seawolf* was moored fast.

A crane lowered the forward brow into position. Men streamed across the brow and into the arms of their families. Sea bags over their shoulders, home they went. Those were forward guys. They shut off their radios and sonar gear with an on/off switch and beat feet like Fred Flintstone.

The Nukes still had work to do. We had to shut down the propulsion plant and the reactor. It would be a couple of hours at least before Nukes were allowed to leave the boat. My first time coming home, my wife had waited on the pier with the other families. I crossed to the pier long enough to hug her, then went back to work. I was free a couple of hours later, but the fun of the event was quashed. What was she to do, wait in the car for hours in a dingy shipyard parking lot?

After that first time, she waited for my telephone call saying I was free to come home. Then she came down to the boat to get me. Sometimes I caught a ride with one of the other guys. Those homecomings were less picturesque, but far more practical.

My tenure as Line Handling Captain over, I climbed down the Engine Room hatch, returning to Nuke Land, a world devoid of ropes and knots. There was work to be done and the quicker we got it done, the quicker Nukes could go to their families. We all pitched in.

25

Top Secret

If someone is compromising the security of the Navy or of the United States, that person is compromising you.

— The Bluejackets' Manual, Twentieth Edition

ONE OF THE first items issued to a fresh recruit at Basic Training was his personal copy of The Bluejackets' Manual. The Bluejackets' Manual was the go-to reference during Basic Training. There it doubled as a clothes iron and a measuring device for bunk inspections. I never saw one used after Basic Training.

The U.S. Navy is serious about the security of its information as well as the physical security of its property. The Bluejackets' Manual, makes the point sublimely: "Unfriendly foreign nations are always interested in classified information on new developments, weapons, techniques and materials, as well as ship and aircraft movements and their operating capabilities."

Navies of other countries likely have the same phrase in their training manuals, except, of course, they are referring to us.

Seawolf was a high-security submarine. Even the cooks had top secret security clearances with special background investigations. A placard in my medical records stated: Notify Naval Intelligence Prior to Placing this Service Member under Anesthesia.

A minor flap ensued when I had a medical procedure done at Travis Air Force Base. That doctor gave me local anesthesia. Afterward, the Doc on the boat noticed me walking oddly. Off to the side, I explained in a low voice about my recent vasectomy. He blanched.

The next thing I knew I was officially answering questions. Apparently Naval Intelligence made no distinction between general and local anesthesia. Who knew?

Lucky me, my quiet little vasectomy became a National Security Issue. It helped not at all that the Air Force doctor doing the deed had entered the operating room wearing a plastic Groucho Marx nose, complete with fake glasses and mustache. Leaning over me he asked if I was ready.

"Doc," I groaned. "What are you trying to do to me?"

He laughed, "You can't be too serious about some things."

"Oh, yes you can, Doc. Yes, you can. This is a perfect time to be serious about some things."

As the procedure began he warned me that I would feel some pressure.

"OK," I said. "I'm ready." Pressure, yes. What an interesting term. During the procedure I promised to buy two boxes of Girl Scout cookies from the doctor's eleven-year-old daughter.

Security clearances were renewed every three years. The office of Naval Intelligence sent a suit down to the boat and one by one we found ourselves answering questions about the shipmate *du jour*. The questions were designed to determine a person's susceptibility to being converted into a spy by the other guys. Maybe he was already playing both sides. The tell-tale signs of those activities could then be sleuthed out.

Although the questions were innocuous enough, the answers could be unintentionally damning. On the flip-side, the person you were being quizzed about was going to be next in the room and they would be answering questions about you.

It was a conundrum. We were constantly reminded not to talk about work, the boat, the schedule of the boat, or anything that may or may not happen on the boat. Now a man I don't know was asking me questions about the people on the boat. I felt a responsibility to be as vague as possible. I phrased every answer to imply the best possible connotations. This was not a treasonous attempt to deceive; my concern was that the Man didn't screw over my buddies because of something taken out of context.

On one such occasion I was being queried about Tom Brazee, who in the past few months had suffered several blows to his well-being. He abruptly divorced and a beloved grandmother had passed away. She did have the grace to leave her grandson a cash inheritance. Newly single with no children, a Navy paycheck every two weeks, and receiving a tidy windfall, he did what most young men did. He bought a new car.

It was a Cadillac Baritz; about as luxurious a car as a young man could stand in those days. We thought it an old man's car and told him so. A Navy bachelor stud should be in a Corvette or a Porsche. He paid cash for the Caddy. Good for him.

Now, Naval Investigative Service questioned me about Tom Brazee.

> NIS: Has Petty Officer Brazee's mood shifted in the past few months?
> ME: He has his good days and he has his bad days.
> NIS: Has Petty Officer Brazee's home life been stable?
> ME: You have to ask him about that.
> NIS: To your knowledge is his marriage good?
> ME: I don't think so. His wife is shacked up with a civilian. I think it may be a guy from NIS.
> (This last bit was not true, but it was payback for the

NIS man making me nervous. The Man wrote in his notebook.)
NIS: Has he recently made significant changes to his lifestyle?
ME: Like what?
NIS: Has he been hanging out in bars more than he used to?
ME: Uh…yeah. A little more. Maybe. I guess.
(The Man wrote in his notebook.)
NIS: Does he seem to live within his means?
ME: I have no idea what his means are. We don't talk about that.
NIS: He's the same pay grade as you with about the same time in service. Does he seem to have a lifestyle similar to yours?
ME: I know he just moved into a really nice apartment and has been buying furniture.
(The Man wrote in his notebook.)
NIS: Has he flashed any money around or made any large purchases lately?
ME: Other than the furniture?
NIS: Other than the furniture.
(I think he wrote about me that time.)
ME: You're asking me if he bought say, a shiny new metallic-blue Cadillac.
NIS: Right.
ME: OK
NIS: OK what? Did he recently buy a shiny new metallic-blue Cadillac?
ME: No, it's red.
(The Man wrote in his notebook.)

Later I ran into Tom doing preventive maintenance on the Primary Plant Instrumentation System in Reactor Compartment Upper Level. He asked how the interview went. I answered honestly.
"It went fine for me, but I think you're going to jail."

We had been at sea for a week or so when a security briefing was held in the Crew's Mess. We were given our cover story in case of capture by people who want to know what the hell a U.S. nuclear submarine was doing in their pond.

The first order of all military personnel is to say nothing. Name, rank, and serial number; that's all they would get from us. However, in the event that they weren't interested in our name, rank, or serial number and decided to play rough we needed some cards to play.

First, Big Lie Number One: We are on a training mission testing out new equipment.

Then Big Lie Number Two. This was a card-up-the-sleeve, to be played only when they were pulling out fingernails. Hopefully it was juicier than Big Lie Number One. With it came the fantastic hope that even though my torturers had not bought into Big Lie Number One, Big Lie Number Two would satisfy them. Sure.

The dream was this: After I laid out Big Lie Number Two in a convincing performance, my inquisitors would say, "Oh, OK. Thanks. You can go now."

We were hoping for this courtesy from the country that took 91,000 German prisoners in Stalingrad. Only 6,000 came home, many not until nine years after the war ended. Nine years after the war. Being an uninvited guest of the Soviet Union would not be a walk in the park.

We didn't know about waterboarding back then, but had we known that waterboarding was the worst that was going to happen to us if we were captured, we would have been thrilled. We wouldn't have bothered with a cover story. We could have just laughed and told them to go to hell.

You're going to do what, trick me into thinking I'm drowning? Ivan, I'm a fucking submarine sailor. I think I'm drowning every time I go to sleep. Bring it.

What worried the Nukes most was our actual lack of knowledge of the mission. We imagined having knowledge that we could withhold would give us a greater sense of value and valor.

"Those lousy bastards didn't get a thing out of me," is what we wanted to growl at the elbow-worn bar in the Old Vets Club.

Knowledge would be currency in our pockets and we had none. We feared that our interrogators, made crazy after weeks of listening to our incessant "I don't know anything!" would shoot us out of plain frustration.

These security briefings were handled in what I thought to be an alarmingly nonchalant manner by the command. Documents telling me what I could and could not say in the presence of our enemy's were typed, briefed, signed and filed away as casually as next week's lunch menu. Later I would discover that 150 pounds of high explosives were purposefully positioned in the boat that would ensure no information, no evidence, and none of us were ever captured. Unless they were printed on Secret Squirrel spy paper that dissolved in seawater, the legal documents I signed would be there on the bottom with us.

The chance that a secret will stay a secret is inversely proportional to the number of people in the know. This is not to say there has to be a bad apple in the crowd for a secret to get into the wrong hands. Keeping secrets is hard work.

A military secret is like a big box of Styrofoam packing peanuts. The secret is not just a benign item, it is information that people use. If a few people reach into the box and work carefully, the packing peanuts will stay in the box. If too many people are grubbing around in the box, packing peanuts end up on the floor. These can be picked up and put back in the box only until someone opens the door. A gust of wind scatters the packing peanuts and, even years later you will find stray packing peanuts every time you move the furniture. So it is with secrets.

To keep an operation secret the first rule is the "need to know." If someone doesn't have a need to know, regardless of rank or security clearance, they don't get to know. Just because I had a Top Secret clearance did not give me access to all Top Secret information in the world. There are different levels of access for any particular operation.

The engineering types whose job it was to get the boat to "wherever," so it can do "whatever," didn't need those details to successfully accomplish their jobs. Some people knew where we were going, but not why. Some knew what we were doing, but not precisely where. The course was plotted behind a curtain with Need to Know Access Only. Even if they did know where and what; most of the crew didn't know why.

Only a select few knew the what, the where, and the why. That's how secrets are kept—theoretically.

In November 1985, we were tied to the pier and struggling with *Seawolf's* obstinate Ship's Service Turbine Generators, trying to ready the old girl for her last mission. The turbines would come up to speed normally but when speed control was placed on automatic, the turbines tripped off. The malfunctioning governors were giving the Engineering Department and the shipyard "experts" fits. This went on for weeks.

The Captain's demeanor in the Engine Room when the freshest repairs or adjustment of the speed governors failed one after the other was icy resolve. "Fix it," was all I ever heard him say. His carriage and demeanor suggested that he was plenty ready to do murder if he could only figure out where the deed would do the most good to get his ship to sea.

One day the machines operated as designed. No one could truthfully give credit to any particular repair or tuning of the governors, on latest take-apart-put-back-together the things worked fine. The shipyard experts huddled to conjure a professionally astute cause for the written report of the malfunction while avoiding the word "gremlins."

During this period of the great Ship's Service Turbine Generator fiasco, a news item caught our attention. One Ronald W. Pelton, a former employee of the National Security Agency, the "huge but super-secret agency assigned to intercept and analyze communications of foreign governments" according to the newspaper, had been arrested and confessed to selling the super secrets of the super-secret agency to the Soviets. According to Washington Post reporter Bob Woodward, Pelton was a low-level staff man with broad access to U.S. intelligence gathering activities pointed in the direction of the Soviets. What the son-of-a-bitch gave to the Russians was us. Pelton told the Soviets about *Seawolf's* and other

DevGroup submarines' missions. The Soviets knew what we were up to, however, they didn't know exactly where.

Pelton claimed the Russians paid him $35,000 for the lowdown on Operation Ivy Bells. The KGB said they paid him $37,000. Somebody pocketed $2,000 that he didn't expect to be missed. That's the trouble with spies; you can't trust any of them. It's like the guy who cheated with another man's wife, and then is shocked, shocked to find she's cheating on him.

On the mission of '86, Nuke Machinist's Mate Petty Officer Second Class Tim Correll brought aboard a cheap cardboard model of the terrestrial globe and stashed it in the engine room. Each day when he came on watch, he reviewed the seawater injection temperatures from the Engine Room Lower Level logs and the RPM of the shafts from the Throttleman's logs.

If sea temperature dropped, he reckoned we had turned north, slightly or radically depending on the magnitude of the change. If temperature was constant, he figured we were sailing along a line of latitude. He calculated distance traveled from averaging shaft revolutions and the pitch of the propellers. Starting Day One at the Golden Gate and extrapolating the additional two parameters, he charted what he thought to be our position on the globe with a fine-tipped marker.

We laughed at the crude effort. We teased him about the enormous lack of information in his process. Ocean currents, ship depth changes, changes in salinity, weather systems, day to night temperature shifts, underwater volcano activity, whale migrations, Sweet Jesus on a whole wheat cracker, you've got to be kidding, Tim…

Correll didn't care. It was his hobby and we could all go straight to hell if we didn't like it.

Fine with us, knock yourself out, moron.

Fuck you guys.

He kept the globe stowed away and we let him be. After a month of amateur course plotting, the Executive Officer caught sight of the globe.

"What's this?" he asked casually.

Correll explained the process. The XO listened patiently but absently, as he would have listened at Thanksgiving dinner to a dotty uncle replaying a trip to the doctor. Suddenly the XO leaned in, looking closely at the black dots across the blue ball. His eyes lost their humor.

"I'm taking this," the XO said.

The XO and the globe went forward to the Control Room. Surprised voices were heard behind the curtain concealing the course plotting table. The XO returned to the Engine Room.

"Petty Officer Correll, I'm confiscating your globe. I am ordering you to not do that anymore."

Correll beamed. "Aye, aye, sir."

Then quieter he said, "XO, how close was I?"

The XO grinned. "If I have to fire the Navigator, you've got the job."

His head high, Correll coolly appraised the rest of us. We had made fun of him. It was now painfully clear; there was nothing he could say that we in the peanut gallery could possibly understand.

Oh brother, this was going to be a long trip.

The Big Lie security briefings were held in two shifts. I managed to be on watch or doing maintenance and missed both briefings. A makeup session was held in the wardroom led by the Navigator, Lieutenant Commander Fredrickson. Machinist Mate Laverne Levelle attended along with me and several others.

Laverne was a nice guy not known for searing intellect. He had no neck, his head blending straightway into wide shoulders. One heavy eyebrow spanned both eyes. With fingers like sausages, his hand looked uncomfortable gripping the tiny government-issue ballpoint pen.

A legal document was placed before me that could surely be used against me in a court of law if I ever wrote a book about all this. Each man in the briefing had one. The document contained separate pages for Big Lie Number One and Big Lie Number Two. My name had already been typed into the appropriate places. Above the blank for my signature was a statement:

I, not under duress, knowingly and willfully agree
to the terms herein.

Laverne squinted at his document. He had a question.

"What's 'duress?'"

The Navigator coolly replied, "Duress is what you are going to be under if you don't sign that paper."

Laverne studied the paper for a long minute. He studied the Navigator who stared back without blinking. Laverne Lavelle scrawled his signature.

"Whatever you say, sir."

Submarine Development Group submarines included a department other boats didn't have. "Projects" was full of serious people who did not talk about their work. Even at sea, sitting in the Crew's Mess having a cuppa Joe after the evening meal, sailors in different departments did not ask about each other's professional business. They might ask when some inoperative piece of gear would be fixed, or if things were going well. But the conversation stayed general. Specifics were out of bounds. One end of the boat did not know what the other end of the boat was doing.

I sat across the table from Joe Flanagan, a good guy but still a Projects Weenie. We each had a cup of black Navy coffee that would float a horseshoe. Joe and I chatted about how we might get together a few guys and go to an Oakland A's game. The season would be half over by the time we got back.

We chatted about more pressing topics. "How's the repair coming on the diesel?" Joe asked. I was a Reactor Operator and didn't work on the diesels, but I was Engineering Department. Most everything aft of frame 57 was engineering department, including the emergency diesels and therefore, fair game to ask a Nuke.

"M-DIV seems to have that under control, they should be done with it soon," I said. "I saw the bad main journal bearing apart. The babbitt bearing surface was wiped slick, right down to the casting."

"Wow," said Joe.

I continued the update. "Yeah, they don't usually re-babbitt bearings underway, so it's been a learning experience. They finally got the bearing resurfaced and now they are putting the diesel back together. It should be up and ready in a day or so."

"That's good," said Joe.

I asked Joe how the mission was going.

"Good," he said.

Still in friendly conversation mode I asked if we were on schedule. Joe confided that they had experienced some glitches but had them sorted out. We were pretty much still on schedule.

"Glitches with what?" I asked.

Joe focused calm blue eyes on my face. His look was the gentle, bemused look a grandfather might give his small grandson when asked why the boy's parents locked the bedroom door when they took an afternoon nap. Joe, like the grandfather, had no intention of revealing details.

Joe grinned. "You know—glitches." Oops, a foul ball on me. But we had been at sea for months. I was hungry for information, and besides, this was a game. It was the submariners' version of Jeopardy. Every answer was a question.

"Are they major glitches, or minor glitches?" This was an acceptably vague question. Informational security likes ambiguity; Joe acquiesced.

"Minor glitches." So ended our conversation on that topic. We moved back into the safer world of generalities and trivia.

Who do you like for the NBA championship, Joe: the Bulls or the Celtics?

Seawolf was fitted with twin propeller shafts, one on each side of the submarine. Each shaft was supported outside the pressure hull by a strut with a bearing positioned just forward of the propeller. Not enclosed, the bearing was "open to sea" on each side. Seawater was sucked into the space between the rotating shaft and the stationary bearing, performing the same function as oil in your car's engine. A rotating "wedge" of water formed which lifted the spinning shaft off the stationary bearing surface. Once the shaft spun fast enough to maintain this hydraulic wedge, the shaft and the bearing no longer touched. The shaft was now spinning on a film of water. This low-tech hydrodynamic bearing had no moving parts and was therefore extremely quiet when operating normally.

Submariners like quiet. Noise put into the sea travels a long way which is bad news in the stealth business.

As only the second nuclear powered submarine, *Seawolf* was built when submarine design was changing by the day. Components that worked well continued as they were until their turn came to be improved.

One thing that continued into *Seawolf* was the traditional material of the shaft bearings outside the pressure hull. Although supported by good American steel, the bearings were surfaced with wood. Lignum vitae, to be exact, a tropical hardwood long appreciated by mariners for hardness and durability. Lignum vitae has been found in marine bearings on shipwrecks dating back to the time of Christ.

When in her operational area and doing her thing, *Seawolf* sometimes traveled very slowly. As in ver-r-r-y slowly. The rotation speed of the shafts was too low for the hydraulic wedge to form between bearing and shaft. The smooth steel of the shaft rubbed against the smooth hardwood of the bearing surface and produced a unique noise; slow, squeaking wails that rose and fell in octaves. If you've heard whales "talking," that's the sound.

For about five minutes this music was fascinating when I first heard it through the hull. It became obnoxious since it could be heard clearly from my bunk in the Stern Room. The starboard outboard shaft bearing was only a few feet from my head. It was outside and I was inside, but the noise was clear enough to keep me awake until I learned to sleep with earplugs jammed into my head.

When the submarine was in this super-slow mode, the joke was this: if Ivan the Russian sonar man was listening, he would only hear whales humping and would pass us by as a biological anomaly.

See there, what I've done. Now when you visit the aquarium and hear recordings of whales singing, you will never be quite sure. Was it really a whale, or was it *Seawolf*?

Sure, I could tell you, but I won't. That information is Top Secret.

26

Man Down

'Ere I break, you'll hear me crack.

— *Captain Ahab of the Pequod*

ON JANUARY 28, 1986 *Seawolf* was resting comfortably in her berth in the Napa River. At about 8:40 a.m., I step-and-ducked through the watertight door into the Crew's Mess to grab a cup of coffee. Men in the Mess seemed frozen in place as they stared at the TV mounted on the bulkhead. On the screen I saw a smoke plume where seconds before had been the Space Shuttle Challenger. Challenger had exploded and broken up 73 seconds after lift-off. The first Teacher in Space, Christa McAuliffe from Concord, New Hampshire, was on that craft. All aboard died. It was a sobering moment for men who went on secret missions in inner space and who were kept alive by a steel vessel made up of millions of parts manufactured and assembled by the lowest bidder.

I recalled my friend at the S1C prototype who had applied for the Teacher in Space program wanting to be the first Navy Nuke in space.

It later was discovered that an O-ring had failed, causing the disaster. A rubber sealing ring. How many O-rings did we have on our submarine? Hundreds, if not thousands. Another sobering thought in a business of sobering thoughts. In our minds, all military contractors' trust stock went down a few notches.

Every three years each nuclear power plant under the jurisdiction of Naval Reactors was required to pass an Operational Reactor Safeguards Examination. If the ship, submarine, or facility passed the examination they continued to operate. If they did not pass, well—let's just say the Devil came to town and set up shop.

Seawolf was due for decommissioning after one last mission. A Ship's Service Turbine Generator problem caused us to be in the shipyard longer than planned. As the weeks slipped by, the window of opportunity closed in our faces. Passing our expiration date, we would have to complete an ORSE before we could go on that final mission. Evidently being a super secret Special Projects submarine readying to conduct a mission of vital importance to the security of the United States did not carry enough weight for Naval Reactors to make an exception. They made the rules and changed them when they saw fit, one would think a small caveat could be applied to waiver our recertification for one last trip. If they were really lucky we might never return saving them the fuss and bother of conducting a *Seawolf* ORSE ever again.

Maybe Naval Reactors had a burr under its saddle blanket for Projects boats. I don't think the Naval Reactors people liked being told where they could and could not go on the boat. The Naval Reactors Regional Office, the acronym for which—NRRO—we pronounced as a word, sounding as N-Roe, were the earthly representatives of the Lord of Naval Nuclear Power, Hyman G. Rickover, himself. A sailor who had been aboard for a Rickover visit in years past told me that the Admiral hated being told he didn't have enough clearance to see certain things on our submarine. Hated with a capital H. Special Projects or not, NRRO cut us no slack.

Even though we Nukes jokingly thought of ourselves as the brighter crayons in the submarine box, we only existed to make sure the Forward guys could do their jobs. We pushed the boat to get them there and kept the lights on while they performed their secret sleights of hand. Nukes were like the guys who set up Pink Floyd's stage and equipment; no one came to the concert to see us, but without us the show didn't happen.

Sharpening the crew's level of knowledge and boardsmanship skills could be done while tied to the pier. The volumes of paperwork that would be audited could be checked and rechecked while in port.

For testing the crews' ability to operate the plant and respond correctly to casualties, there was no substitute for the benefits of taking the ship to sea. Sure, you could simulate the reactor unexpectedly shutting down while simulating full power. You could also actually shut the damn thing down while answering a flank bell. When the reactor plant alarms are clanging, and the warning lights are flashing for real, and the propulsion plant is shutting down around you, it produces a unique sensation in your gut. Hearing rotating machinery and electrical loads winding down and going quiet, gets your attention in a way no amount of simulation could replicate. Feeling the boat slow as the propellers stop adds real-time urgency. There's "Theory of Operation" and then there's "Operation."

The crew was well trained and each member had demonstrated his understanding and ability to make correct decisions in daily operations as well as in urgent times. These abilities had been proven, documented, and certified. But weeks and months in the shipyard will rust even the sharpest edge.

Sitting static and impotent alongside the pier, the submarine grew a film of living algae and dead microbes out of the organic brew of the green river water. The boat needed the sea to scrub the crud off her hull. The crew dulled by cold-iron life also needed a good dusting and stretching-out to make them fit, trim, and seaworthy. Remember how your grandmother hung a rug over the clothesline and whacked it with a broom until no more dust came out? Training a submarine crew was much the same and frighteningly similar, in a metaphorical kind of way. So—off with the moorings and out to the big water.

We cycled through routine operations and non-routine operations. We ran casualty drills and special evolutions on every watch section. Only a few hours were uninterrupted enough to get any sleep. We were tired. Tired and stressed from a week of adrenaline under the heavy thumb of perpetual catastrophe.

Immersion in stressful situations, where a dozen things are happening at once, is how a person becomes accustomed to quickly organizing his thoughts in the midst of chaos. Alarms ringing, warnings buzzing, reports streaming in, and orders flying out overwhelms new trainees. Even an experienced man can lose track of the goings on as well as his composure. A fundamental in-your-bones understanding of how and why everything works is vital.

In many businesses, "why" is not the purview of front-line employees, only of senior management. In submarines, why is paramount. Why is the key to what and when.

After you have drilled a scenario, debriefed it, analyzed it, discussed it, learned from it, and trained on it, when it happens for real, peoples' actions fall in line. That's the plan. I've been here before; I know what to do. Sometimes it works smoothly, but catastrophes tend to make up their own rules. Knowing the fundamentals backward and forward could make a near-disaster seem routine.

Continuous stress manifests itself in ways that you expect, but also in ways you do not expect. I watched a man fall asleep while standing up with a steaming mug of coffee in his hand. We all had been up and at it for thirty six hours. Several crew members in Engine Room Upper Level were discussing the previous drill. For a moment the talk fell silent and Brett's eyes drifted shut. He began to tilt and as his hand sagged, one man lightly swept the coffee cup out of his hand. A hand on his shoulder prevented Brett from tipping over completely.

Brett's head jerked up as he came awake with a start. His coffee cup was handed to him and the discussion continued without a hitch. No one mentioned his momentary lapse of consciousness. No one seemed to notice. That was the kind of thing to be expected during an underway ORSE workup. Exhaustion made weirdness routine.

When the Engineer Officer stepped up to Skip Lethin, the Engineering Watch Supervisor, and informed him that he, the Engineer, had just drunk hydrochloric acid, the bleary-eyed Engineering Watch Supervisor chuckled.

"Yeah, right," said Skip wearily. "Good one, ENG. I may do that myself before this ORSE is over."

"Chief, I'm serious. I just drank the hydrochloric acid from the Secondary Sample Sink."

Chief Electricians Mate Lethin stared, gaging the man's sincerity while his own head spun with the implications of the unexpected confession. No drill had prepared him for this moment.

A two-ounce bottle of hydrochloric acid was stored in the Secondary Sample Sink. One of the steam plant water analyses used a few drops of the acid to produce a chemical reaction. The Engineer had apparently tossed back the contents of the half-full bottle like a shot of tequila.

What in the hell kind of damage would hydrochloric acid down the gullet do to a man? Chief Lethin didn't know the medical details, but he knew that it could not be good.

The Chief's head spun. This was an emergency, yes, but it was also a very personal emergency. Should he deal with this quietly? Should the crew know that a high-up member of their command had lost it? The man standing before him seemed fine at first glance. Was it really even an emergency? After a moment of consternation, the Chief decided. He didn't know exactly what kind of situation faced him, but he knew what to do when he didn't know. He called away the injured man as he would any other casualty. Help would come.

When bullets are whizzing on the battlefield and a Marine yells "Medic!" it's a Navy hospital corpsman who comes running. Within the space of a few deep breaths, *Seawolf's* Doc was on the scene.

Doc took charge. He shot rapid-fire questions at the Engineer. What did you ingest? How much? How long ago? Doc probed and poked the ENG, lifted his eyelids and peered down his throat.

Doc listened to the ENG's heart and took his pulse. He never asked "Why?" of the Engineer. At the moment, that answer was irrelevant.

The Engineer Officer was a Lieutenant Commander. He outranked the Senior Chief Hospital Corpsman every day, if a duel of collar devices had been appropriate. At the moment, pay grade was immaterial. Regardless of rank, the LCDR himself was Navy equipment assigned to the Doc. Responsibility trumps rank, and the urgency of the situation trumped military courtesy.

Doc would have balanced the scale at about 220, all of it packed solidly on a six-foot four-inch frame. The crowd in the passageway parted when Doc ordered, "Make a hole!" Doc carried the 170-pound Engineer under his arm like a child to the sink in the aft shower room. The shower room was barely big enough for both men. Holding the Engineer head down toward the stainless steel sink, Doc jammed his finger down the man's throat, inducing him to vomit.

The Engineering watchstanders didn't know what to do. It was in our natures to go toward a casualty, to get in there and find out what was going on and to take action to minimize damage and recover the plant. But this was different.

This was an intensely personal casualty, way out of our depth. We felt bad for the Engineer. We were also dazed. This was our department head, the number one man in the Engineering Department, the man who gave final approval to everything we did. Now he was laid low.

We couldn't get a handle on the situation. We didn't know jack about how to help someone who had slipped into the void. Every man there would have thrown the ENG a lifeline, even at his own peril, but what to do for this?

Jesus, I thought. Do you suppose this happened because he was exhausted? Maybe his brain fell asleep and mixed up signals. I hadn't slept more than a few minutes here and there in the past two days. As the guy who would be held responsible for the whole department, I doubted the ENG had slept more than the rest of us. Probably he had not slept at all for at least the past two days or longer. The pressure placed on the Engineer Officer for the boat to do well in the ORSE was immense.

Milk from the crew's mess arrived. A routine of drinking milk and throwing it back up began and continued. Doc was going to get the ENG's system flushed out.

The ship had been making its way back toward the Golden Gate at a leisurely pace, drilling and practicing maneuvers in safe waters as we sailed toward San Francisco Bay.

Now a flank bell was rung up. We put spurs to the horses and the engine room roared to full power. High pressure steam poured into the Main Engines, spinning the turbines at maximum RPM. We hauled ass for home.

The Doc had done all he could. The acid that could be removed from the ENG's system was out. The rest would be up to doctors with more equipment. The ENG rested in his stateroom with the Doc attending while we cut a wide wake toward San Francisco Bay.

A Navy helicopter met us a few miles outside the Golden Gate. The Engineer rode a sling up and disappeared into the helicopter. That was the last most of us ever saw of him.

He was a likable man, a good Engineer, and above all; he was our shipmate. The crew was shocked by the event. We were also curious. This curiosity carried no morbid intent, no horror movie reflex compelling us to stare at the axe in the head of the camper gone missing. We were keenly aware that we were only as normal as he. We had all spent time underwater together. What had happened to him that he decided to do this?

For a week the Nukes discussed who said what to the Engineer trying to figure out what had set him off. How jangled do a man's nerves have to get before...? No one could finish the thought, it was too confusing.

Stanton, you saw him ten minutes before he did it. What did he say? Was he acting weird?

How long do you suppose he thought about it before he did it? A week? A few minutes? Three seconds? Did he do it and then think, "Why did I do that?"

We were unaware of what was going on with the Engineer now that he had been whisked away. We didn't want to pry into the man's personal business, but we did want to know. Was he dead, would he be dead, could they fix him up? Would he come back to the 'Wolf, would he still be in the Navy? The whole situation was new territory.

Forward guys came back aft and asked about the ENG. We had no answers. I was miffed when a forward guy asked, "What the hell did he do that for?" Never mind that was a question to which I did not know the answer, it felt as if an outsider was prying into family business.

Not knowing gave rise to speculations of all sorts. The ENG would be court-martialed wouldn't he? He would be stripped of his commission; he would probably be fired from the Navy. Certainly his security clearance had to go away. That would disqualify him as an Engineer—or anything else for that matter—on a nuclear submarine, anywhere, anytime. Professionally he was surely done in the Navy.

In the civilian world these idle suppositions may be called water cooler chatter. In Navy parlance it was "sea-lawyering." Experts sprouted out of the deck plates.

The wiser among us refused to offer an opinion, declining to speak on issues of which they were unknowledgeable. Some Engineering department members were angry. We had been working up to an Operational Reactor Safeguards Examination. An ORSE was a bitch, big time, and it was the Nukes that would be in the spotlight. Now, a week before the big event, the Engineer—the Head Nuke—had bailed on us. Concerns about the ORSE were mostly shrugged off.

It's on us, said the more experienced of the crew. The ORSE tests all of us. Whether the Engineer was with us or not, it's still on us to pass the ORSE. And we will pass the ORSE.

Even though most everyone on the boat was sympathetic to the Engineers misfortune, sympathy did not prevent rude jokes from making the rounds. After all, we were sailors. Making jokes about things not funny was in our DNA.

Most of the quips related to hydrochloric acid in shot glasses and killer cocktails. In defense of our lack of decency, the jokes were aimed at the event, not at the man.

Anytime someone went outboard in Engine Room Upper Level where the Secondary Sample Sink was, a comment was sure to follow them.

"Hey now, what are you doing over there? It's not happy hour at the sample sink!" Having this line tossed his way, the Executive Officer responded in kind.

"You don't have to worry about me," the XO said. "Suicide is not an option for me, but homicide is definitely on the table."

It turned out that *Seawolf's* Engineer survived the incident and is currently enjoying a successful life. I'm glad for that; I wish him well. I imagine he spent a lot of time trying to recall what was in his mind when drinking hydrochloric acid seemed like the right thing to do.

I wonder if he found an answer.

27

ORSE

Life's a bitch; and then you die.

— *The Submarine Sailors' Imaginary Book of All Knowledge*

BEFORE *SEAWOLF* COULD leave Mare Island to embark on its final mission before decommissioning, it had to be certified by Naval Reactors as a safe nuclear operator. The certification process was well known to the Nukes. It was officially an Operational Reactor Safeguards Examination. We called it ORSE and it sucked.

Writing this, I asked my old submarine buddy Dennis Winn what came to his mind when thinking about an ORSE. His response: "Fear. And a shitload of work."

ORSE was a make or break deal. If the command passed the examination, great. Terrific, good for us, time for a party. However, if we failed the ORSE—every Nuke shuddered at the thought. We would have to remediate ourselves until we passed a repeat ORSE. The repeat exam would be scheduled several weeks later.

In between would be hell. Remediation would be brutal. The command would impose a harsh work load on us to declare its commitment and to properly motivate its deficient crew.

Fail became the F-word and not allowed in the engineering spaces. Saying it aloud would plant unwanted seeds in the garden of the Fates.

We had an additional challenge to deal with during *Seawolf's* last ORSE. We had a new Engineer Officer. Our new ENG had been called up with literally a moment's notice. "*Seawolf* is minus-one on Engineers. Pack your sea bag" may have been how that conversation went.

The new Engineer Officer had a tough job right in his face. He was to take over a submarine engineering department without the benefit of a formal turn-over process from the outgoing Engineer. Not only did he have to learn the 575's unique systems quickly, he also had an ORSE on his calendar much sooner than possibly anyone in history.

If anyone could be up to the job, it was our new Engineer. Faced with the time challenge he was under, it would have been stereotypically easy for the new man to come in with a hard-charging, here's-how-we're-gonna-do-it attitude and whip the department into shape.

Many Navy officers put on a show of importance. They carried themselves as hard men, all-knowing and uncompromising. I suppose they had been told that a perpetually furrowed brow would make them appear as if they held together the entire command by the power of their will. Perhaps they believed their path to respect was the image of importance. I wondered if they knew how everyone rolled their eyes and shook their heads at the display.

The world has an endless supply of assholes. An inordinately high percentage of them precipitate out of the primordial stew and settle to the bottom of the mixing bowl that is the world's militaries.

Also in those same militaries will be some of the best people you could ever hope to meet. Of such was our new Engineer.

It would have been understandable if our new Engineer Officer had jumped into *Seawolf's* Engineering Department with a "my way or the highway" attitude. He had a lot to do and no time to do it in. Instead, our new ENG came in quietly. He asked lots of questions from everyone, even the men lowest in the engineering department. He listened intently and asked clarifying questions. He asked good questions.

He didn't need to prove his competence. Competence, both as an engineer and as a human being, was written all over the man. He didn't have to declare himself as in charge. He was as easygoing as most of the crew. This is not to say we weren't serious about the work. There is a time and place for the utmost gravity, and a time for a more lighthearted approach.

We saw the way the new ENG's eyes crinkled at the corners when someone made a rare good joke. We saw that he caught the wit of the banter sailors tossed around like a verbal volleyball.

The 575 was filled with competent, qualified sailors. Most of us had been ORSE'd before. We knew our jobs. We didn't need a better system, we didn't need more training, and we didn't need more discipline. What we needed was a leader who we knew had our backs. Our new Engineer was just that man. We trusted him. When he made a decision, we were all on board because we had seen the man learn the plant and learn us. There was no bullshit bravado about him. He talked straight and he made sense when he talked. We trusted him.

One day I remarked that no one referred to him as the "new" Engineer any more. The Electrical Operator shrugged.

"He's one of us now. Whether that makes him famous or ruins him; he is one of us."

Anything that might favorably impress the soon-to-arrive ORSE board was done. Each enlisted man was assigned an area within his divisional spaces for painting. Pipes, motors, pumps, the bilge areas, the bulkheads; fresh paint gleamed everywhere. Of course, there was the requisite flap over the authorized shades of white, dark gray, and light gray. Did the light gray have too much blue? A section of Engine Room Lower Level was repainted to cover that goof. Shipyard guys gave us the new paint and said "sorry 'bout that" as they went home on Friday. Navy guys repainted Engine Room Lower Level while the yardbirds were off for the weekend.

New linoleum was laid on every passageway and walking surface in the boat. Yardbirds did this work. Navy guys painted around them and repainted where the yardbirds scuffed the paint.

All light bulbs were changed out to make sure all areas were as bright as possible. Bright and shiny was our mantra. New watertight covers were installed on the fluorescent light fixtures, even if the current cover was still good. We wanted uniformity, all light covers must match.

The boat had a dozen or so sets of the multiple-volume Reactor Plant Manual. Each of these books weighed fifteen pounds. If one fell on your head it would leave a permanent dent. The spine of the book was metal, heavy black covers were hinged to the spine. Replacing its waterproof pages required a screwdriver . The Reactor Plant Manual was an industrial quality set of books. I had no idea how much those book covers cost, but we replaced them all with new ones even though the boat, and the books, were scheduled for decommissioning after this trip. The manuals on the bookshelves in every engineering space were uniform, bright, and shiny.

Sailors wearing faded uniforms were instructed to get new uniforms before ORSE. Personnel inspections were held to make sure every person was as uniform, bright, and shiny as possible.

"You need a haircut, Thompson. Gerwitz, shine your goddamn shoes! You been working in a barnyard? Davis, D.—not you, Davis, C.—get a new rating badge on that sleeve. That one is curling at the edge."

"Now commence Field Day," crackled over the ships announcing system. A chorus of groans rose up as if lead by a choirmaster.

I was in the Log Room on one occasion field day was called. The Log Room was the official boat library of manuals, machinery records, old watch logs, and the like. The Log Room was about the size of a cheap apartment closet and was located almost as far back in the boat as one could get. I was doing official paperwork and hoped I might go through the field day unnoticed. No such luck. The log room petty officer—a collateral duty—was my sea dad, Bob Cranker. He burst into the log room while I was making machinery history entries—one of my collateral duties.

"Rub-a-dub-dub, clean up the sub!" he said. Even when it sucked, Cranker was cheery. I told him that field day could kiss my skinny white ass. Undaunted, Cranker laughed at me.

"Come on you lazy bastard. We can't kill Commies for Christ with a dirty boat!" I have always admired a well-reasoned argument. Alright, alright already, I'm on my way.

Machinist Mate First Class John Parrott was cleaning in engine room upper level. One of the passageway deck plates was lifted. John, uncomfortably perched on the pipes below deck level, wore a blue kerchief on his head. Busy cleaning, his dew rag slipped down over his eyes. His hands full, he peeked under the kerchief and saw several pairs of shoes.

"Hey fellas, can I get a little help with that dustpan?" he said indicating the nearby black metal pan just out of his reach. The pair of shoes nearest took a step toward the dustpan.

"And that foxtail, too," John added.

The handle of the foxtail descended into his grasp. The dustpan was laid directly in front of the pile of dirt and debris John had gathered.

"Hold that right there," John ordered as he carefully swept the pile into the dustpan. He canted up his head to look at the person kneeling in front of him holding the dustpan. Parrott was speechless for the only ten seconds of his entire life.

"Petty Officer Parrott," the Captain said mildly, still holding the dustpan in place. "Do you see anything wrong with this picture?"

Recovering his power of speech and his imperturbable nature, John Parrott grinned and replied, "Lots and lots of things, Sir." Parrott reached for the handle of the dustpan," I'll take that now… thank you sir…really, I got it…thank you…appreciate the help… thank you…"

In the final minutes of the typical four hour cleaning period the deck was waxed. Half of the walkway, an imaginary line down the middle, would be given a layer of liquid wax with a sponge. Lives would be threatened as people needed to pass along the wax-wet deck. The dry side represented life; the wet side represented death… horrible and excruciating death. Unless—of course—the person passing by was the Captain. Then we said "Yes Sir, go right ahead" and left out the threats of horrible and excruciating death. If, however, in his hurry to be about the business of the ship, the Captain inadvertently stepped into wet wax, the Petty Officer in Charge of Wax would say something to the effect of: "Thank you, sir, for showing me which side needs to be redone."

If a Chief Petty Officer made the same slip, he would likely borrow the waxy sponge and apologetically wax over his own footprint. He didn't have to, but he would. The good chiefs would. A shithead would just go on his merry way without a word because he was a Chief and you weren't.

When the first side of the passageway had dried, the second half would be waxed. The same threats would be issued to people now passing on the dry side.

If you have been following along in this entire conversation about preparing for the ORSE, you may wonder what all of the above had to do with being capable of safely handling the nuclear reactor. Ah, ha! Well done. Good question.

The preceding events; cleaning and painting, renewing and polishing, shall we say, the aesthetic rejuvenation of man and machine, were laudable efforts. In the business world this is known as branding. Give the company, or the product, or the person we are selling, that certain *je ne sais quoi*. These physical appearance upgrades would attach positive attributes to the ship and crew when the Naval Reactors team came aboard to administer the ORSE. Even though they were smart guys, the ORSE team couldn't help but get the feeling that we were squared-away and competent.

In theory we were aiming for perfection in all things nuclear, and Navy. However the ORSE team would be looking for faults. One crafty old Master Chief Electronics Technician, the bull Nuke, was a Reactor Controls man. He instructed the petty officer in charge of the RC-DIV Preventive Maintenance Schedule to make up a new PMS schedule and misapply the required color codes to a couple of items.

"But Master Chief, why would I intentionally screw it up?" the petty officer asked incredulously.

"It's these guys' job to find errors. If we give them a couple they'll feel good and quit digging," said the Master Chief. Aye, aye, boss. In business, this would have been us anticipating the needs of our client.

Topside on the pier when the ORSE team came on board. I stood aside to let the group of khaki-clad officers from Naval Reactors cross the brow in front of me. This was the courteous thing to do. Also I wanted to get a look at the bastards. I hoped to gain some sense of character or personality that might help the engineering department. I was looking for a tell of them being nice guys or affirmation that they were indeed humorless pricks and likely to be tough on us.

The ORSE team was pretty much regular people; professional, courteous, and well-mannered. They greeted the Officer of the Day warmly and exchanged pleasantries.

As the ORSE team passed near me to step up onto the brow, I stood at attention and saluted them as a group. One of them, a full-bird Navy Captain, stopped in front of me. He returned my salute and looked at the name on my shirt.

"Are you related to Rear Admiral Heckman?" he asked. His eyes shone warm with the possibility of a mutual acquaintance.

This question caught me off guard. I was surprised that he stopped and spoke to me in the first place, more so that he spoke in a casual, unofficial tone, man to man instead of Navy Captain to Petty Officer. Even so, I would have expected him to say almost anything to me other than to ask if I was related to a Rear Admiral. I had never heard of Rear Admiral Heckman.

"Well…uh…no sir," I stammered. I'm not sure what inspired me to so it, but I leaned slightly toward the captain.

"I really can't talk about that, sir," I said with a conspiratorial smile and what I hoped was a twinkle in my eye. The captain nodded knowingly.

"Of course," he said. He smiled at me and continued on his way.

It's possible that the ORSE team captain now believed that I was related to his friend Admiral Heckman. How could I decide what the man thought? Whatever decision he had made about me, he had made on his own, right? I had told him I didn't know Admiral Heckman. My conscience was clear. Not crystal clear—but clear enough.

The ORSE team aboard, we put *Seawolf* to sea. We stayed surfaced for a longer period that normal on going out that day. Perhaps another submarine was operating in the area and submerging was undesirable. Maybe it was a ploy to make the ORSE board crazy. Whatever the cause, we stayed on the surface for several hours.

The seas were rough and the boat tilted and rolled like a roller coaster. At one point *Seawolf* rolled forty-four degrees to port. Then of course, the boat rolled nearly that much to starboard. Things we thought were rigged for sea came crashing onto the deck.

Three bolts sheared that held a fire extinguisher to the deck. The thirty-five pound fire extinguisher and its holder bounced around the compartment like a beach ball. Catching it was a dangerous proposition. Not stopping it was even more dangerous. One of the machinist mates grabbed a foul weather jacket and charged the rogue extinguisher like a matador. Once the fire extinguisher was caught up in the jacket, getting hands on it was a much safer proposition.

A drawer in an overhead stowage locker fell out of its slide and struck Chief Warrant Officer Wiseman on the head. His were not the only stitches the Doc put in that day. Someone forward had gashed his thigh on a sharp corner as the boat rolled. Doc sewing them up with the boat pitching and heaving might have been worse than the injury. With the boat moving like it was, it's likely that a few false starts were made with the needle.

The Diesel Workbench was an imposing metal structure roughly six feet long by two feet wide located just starboard of the passageway at the forward end of Engine Room Upper Level. It had drawers at each end containing tools and associated paraphernalia for maintenance and repair of the diesel engines. The two massive diesel engines were backup, in case our nuclear propulsion system went completely FUBAR; we could still get home with the lights on. We carried 16,000 gallons of diesel fuel which theoretically was enough to get us to safe haven from anywhere in the world. Theoretically.

The diesel fuel contributed to the smell peculiar to submarines, even though the fuel oil was in a tank. Even though our air was filtered and cleaned, when you live in a watertight environment every odor becomes part of the tapestry and a fart is immortal.

Hanging on to the Diesel Workbench in the rough sea, Machinist Mate First Class Rob Kersch whistled to get my attention. He pointed to the deck. The outline of the heavy diesel workbench was clearly pressed into the new linoleum. The workbench had no legs, its bottom rested squarely on the deck like a box. A clean edge of dried wax was visible where the deck had been waxed around the workbench before we went to sea. With rough seas bouncing us around, the Diesel Workbench had moved.

I yelled at Rob. "I thought this thing was bolted down!"

Rob grinned. "So did I!" he laughed.

I was not prone to seasickness, but today I made an exception. Everyone was seasick. People not on watch lay down and tried to stay in their bunks. Keith Lyly turned a belt over a pipe in the overhead above his top-of-the-stack bunk. With his arm twisted through the belt he stayed in his rack, but surely didn't sleep.

On watch as Reactor Technician, I monitored equipment in Reactor Compartment Upper Level and took two pages of logs every half hour. When the submarine was rolling hard like this, the best place to be was centerline as low in the ship as possible. The reactor compartment had three steps down just forward of the engine room watertight door. I sat on the steps and turned the six-inch-round air conditioning vent so that it blasted air straight onto my face. I was green around the gills, but the cool air kept me alive.

The Engineering Watch Supervisor came in and sat next to me on the steps. He wasn't in great shape himself. He reviewed my logs, initialing each line. "You're late on your 1030 set of logs." Yes, I acknowledged. I didn't move. He didn't move. After a few more stomach churning rolls of the ship, he lurched upright with a forward pitch of the boat. "Catch up your logs as soon as you can," he said. Aye, aye, Chief. He disappeared aft. I remained seated.

Finally we submerged. As the boat smoothed out, so did the men. Those who weren't needed elsewhere slept. People on watch relaxed into the seats at their watchstations. The cooks could finally put together a meal. Everyone breathed deeply and began to feel better. We heard we were granted a couple of hours respite before the ORSE would start.

I thought it grand of the ORSE board to give us a break like that until I was set straight by Machinist Mate Chief George Short. He had just been up to the wardroom where the ORSE team was passing the time before starting the ORSE.

"It looks like a hand grenade went off in there," George said. "Dead bodies everywhere." My brain was still sloshing side to side. I did not catch the metaphor.

"What the hell are you talking about?"

"The ORSE board. They're sick as dogs and laid out on every horizontal surface. Just like you RC-DIV weenies after too many Shirley Temples at the Horse and Cow. We won't be starting the ORSE for a while, I guarantee you that."

See there—I thought the ORSE guys were being generous.

After a few healing hours the ORSE was conducted. The examination lasted three days. Three very long days. An ORSE event was happening in every one of those seventy-two hours.

The engineering department completed written tests, oral examination boards, routine plant evolutions, non-routine plant evolutions, and casualty drills. The three watch sections were drilled separately. After my section was relieved of the watch, we became the Casualty Assistance Team for the on-watch section. After that came interviews on a myriad of subjects. Some interviews related directly to the man's divisional specialty, some interview topics were in a subject not his specialty. No one in the department got more than a couple of hours sleep during the ORSE.

Seawolf's uniqueness worked in our favor. Common in the fleet was the S5W reactor plant. A typical S5W boat's electric plant had five electrical generating machines sharing the entire electrical load of its submarine. However, *Seawolf* did not have a typical electric plant. *Seawolf's* S2Wa reactor plant was the only one of its kind. Instead of five machines powering the electrical grid; the 'Wolf had seven. Seven machines operating in parallel that could be cross-connected in many ways untypical of a more typical plant.

The ORSE board initiated a drill that dropped an electrical buss. Instead of the flurry of activity they expected, the Electrical Operator shut one breaker on the Electric Plant Control Panel. End of problem. The ORSE board monitor in Maneuvering was surprised.

"You can't do that!" he said in amazement.

Sure, said the Electrical Operator. Look, it works like so. The drill monitor left Maneuvering to confer with his cohorts. A few minutes later he came back.

"OK, fine. So you can do that," he said with a hint of sheepishness. As a rule, ORSE board members did not show signs of humanity.

The ORSE team captain that I had spoken to on the brow always nodded to me in passing. Each time he did, I innocently returned the nod. Surely he was just being courteous to a fellow submariner. I couldn't think of any other reason for his being friendly to me.

On returning to Mare Island, the ORSE board went away to tabulate our performance and write the report. The Engineering Department, and most of the forward guys and the Projects guys, celebrated the "End of the Damn Thing" at the Horse and Cow. A week later we received the good news that *Seawolf* had passed her final Operational Reactor Safeguards Examination. Everyone was happy; the Engineering Department was thrilled. We adjourned to the Horse and Cow and celebrated "Not Having to do the Damn Thing Again."

28

Sea Time

Men go back to the mountains, as they go back to sailing ships at sea, because in the mountains and on the sea they must face up, as did men of another age, to the challenge of nature. Modern man lives in a highly synthetic kind of existence. He specializes in this and that. Rarely does he test all his powers or find himself whole. But in the hills and on the water the character of a man comes out.

— *Abram T. Collier*

THE CREWS' MESS on a submarine was more than a place to eat. It was a community space. Here was held trainings, ceremonies, meetings, and entertainments. The entertainments were marginal at best, but enthusiastically welcomed. You might come to the mess decks for a cup of coffee before taking an evening watch and find the place darkened with a movie showing. After breakfast was cleared away you might find Department or Divisional Training taking place. At any time, you would likely see off-watch men quietly reading, their minds in a faraway place lit with sunshine.

I read hunting and fishing magazines shared with Rob Kersch and Keith Lyly. I would bore into an article until the submarine faded from my consciousness. Jack O'Connor had written the words on the page; but in my mind it was me on that high ridge in Alaska. It was me, glassing for a Dall ram on those slopes above timberline. I could feel the chill of October on my face. I felt the weight of the binoculars balanced on my fingertips. I was aware of the heft of the pack on my back and the Model 70 Winchester slung over my right shoulder. Too soon I had to come down out of those hills. I handed the rifle to Jack, closed Outdoor Life, and went aft to relieve the watch.

In the crew's mess you could get into a game of cribbage or a chess match. Depending on which division your opponent belonged to, you might have to settle for a game of checkers. Hearts was a favorite card game around the chief's table.

Men relaxed in the crew's mess so long as their work was done and they were qualified in submarines. Those who had not yet earned their dolphins would not dare be caught in any form of relaxation or recreation. They would be rebuked by all passersby for being "Non-qual pukes," and breathing undeserved air that should only be used by real submariners.

"Get hot! Get qualified!" would blast down on a non-qual every time he slowed down. Non-quals were not authorized to enjoy life. Non-quals were authorized to get their lazy asses qualified. They were allowed to eat, sleep, and breathe as a generosity and only as an aide in their quest to have "Qualified in Submarines" typed into their service records. Only then would they be entitled to wear submarine warfare insignia—dolphins—and sign "(SS)" with their rate and rating.

At sea, instead of an alarm clock rattling me awake, a low voice called my name from outside my bunk curtain. Out of courtesy, the guy doing wake-ups was trying to wake only me, but I had heard him wake other men. Surely other men were roused when he called for me. If I didn't respond to his hoarse whisper, he shook me. The shake was mostly a quick shove on the shoulder. The guy waking me didn't really want to reach into my rack and touch me. Grabbing a sleeping man was dangerous.

Machinist's Mate Tony Hill was famous for taking a swing at the man waking him and later denying having done such a thing. He slept so soundly that calling his name from outside his bunk curtain had no effect. The second trick was to slip open his curtain hoping the sound of the curtain hangers sliding in their aluminum track would wake him. That woke me every time. A nice thought, but it never worked with Tony. You finally had to reach in and shake him roughly. The job usually required more than one shaking. Then Tony would awake with a start and swing a fist at the man waking him. Waking Tony was fraught with peril.

He slept so soundly that we had a joke about it. "If we've been fighting a hydraulic fire for hours, and the battery blows up, and then we get hit with a Russian torpedo, before we go to the bottom the last man out of the boat should try one last time to wake Tony Hill."

It was always night in the berthing spaces. Red lights glowed enough to cast eerie shadows that made discerning detail a fifty-fifty proposition. On any given day, when I woke I rolled out of the rack and squirmed into my blue poopie suit. On a good day, I could put a foot through the pants leg and straight into my shoe, one after the other. But every now and again I jammed my leg down the wrong side of the one-piece coverall. After all, I was still seven-eighths asleep and it was dark. Anyone watching would have snickered at the sight of me turning round and round trying to get myself sorted out while my fuddled mind wrestled in the dark to understand why my legs were not working right. "What the hell…?"

Submarine life aged a man prematurely. The artificial environment, perpetual fluorescent lights, and round the clock schedule took its toll on the human body. A "normal" night's sleep from say, ten at night to six in the morning, was unheard of on a submarine at sea. You might as well get over it. It wasn't going to happen.

On an occasion when *Seawolf* was in port and the people senior to me in my division were on vacation or off to schools; I became

"Acting RC Division Leading Petty Officer." One of my Leading Petty Officer duties was to conduct a "Welcome Aboard" interview for Reactor Operators newly reported to the ship. The purpose of this interview was just as it sounds. It smoothed the process of getting the new man settled in and productive as soon as possible. As part of the spiel I asked his career intentions.

If the new man was planning a career in the Navy, I suggested ways to increase his advancement probability. If he wanted to do his time and get out, I suggested schools we could send him to that would benefit the division as well as add to his civilian resume. Most guys declared themselves undecided. This was a savvy response even if untrue. The man's self-preservation genes avoided insulting any yet unidentified lifers that he may be working with, or for, or even speaking to.

The undecided tactic also kept his options open and invited the U. S. Navy to make him an offer he would not be able to refuse. A desired duty station, sought after schools; anything you want Petty Officer On-the-fence, let's make a deal, was his hope. Astute as the ploy was, it was chronically ineffective. Uncle Sam's Navy did not bargain with individuals already swinging on the hook. Offering Don Corleone "an offer he cannot refuse," always ended badly for the would-be dealmaker.

On the occasion of giving this interview to an undecided new guy it would have been unfair of me to sway him in any direction based on my views. It was his life, I had no business influencing him. But it was also my duty as a fellow human to enlighten him as I had been enlightened. To give the man decision making food for thought, I innocently asked Petty Officer Undecided to guess the age of a passer-by. After observing Senior Chief Gillespie, the Bull Nuke, the newbie took a shot.

"He looks about my Dad's age—I'll say he's 55."

On hearing that the Senior Chief celebrated his thirty-eighth birthday only the day before, Petty Officer Undecided's eyes glazed over. I waited as he stared sightlessly at the bulkhead behind me, his mind ticking off bullet-points leading to the undeniable conclusion. Submarine life was not normal.

At that moment a blueshirt came into the space. His birthday was near mine and I kept rolling with the age theme.

"Joe here is turning eighty-three next week," I said to the new man. It was untrue; Joe would be twenty-six.

"Fuck off scumbag, and fuck you too, new guy."

I explained that Joe was the SNOB, the Shortest Nuke on the Boat. He was chronically cantankerous and planned to remain so until his End of Active Obligated Service. Just prior to EAOS, Joe would have a ceremony at the Horse & Cow to officially turn over the office of the SNOB to the ASNOB, the Almost Shortest Nuke on the Boat. It was a title of no distinction and very little respect. Petty Officer Undecided came to another conclusion. Submariners are not normal.

At sea the only sense of time was the clock. Perpetual light in work spaces and near-darkness in berthing spaces confounded a sub sailor's internal chronometer. I don't know much about Circadian rhythms, but I know that working around the clock and continuously shifting that schedule worked a man like a threshing machine works a stalk of wheat.

The visual cues of daylight, twilight, or nighttime did not exist. There was no sense of the rest of the world. I awoke and started my day. Time was relevant only as a mutually agreed-on point to begin and end my watch.

Meals were a time marker. Four meals were served each day on a traditional rotation, breakfast, lunch, dinner, and midnight rations—MidRats. Which meal was being served when I woke gave me a sense of what was happening in the world outside the hull of the boat.

People get strange at sea. This phenomenon had the same characteristics as alcohol intoxication. I think it erroneous to say people "weren't themselves" when drunk. I think that when drunk, people lose their inhibitions and their true self becomes plain. Seemingly nice guys normally maintaining a good facade become cantankerous bastards when drunk because that's who they really are. So it was with time at sea.

The quiet get quieter, the loud get louder. The extroverted giggle at everything. The stoic slowly crumble into dust and rubble.

George Short was a newly-minted Chief Petty Officer. I knew him back when he wore blue shirts instead of khaki. A member of M-Division, George was a stand-up guy. At one point he had become more rotund than the Navy preferred. George got serious about aerobics and dropped forty pounds. For a year his re-tailored uniforms fit well for a time and then hung on him like burlap sacks. Another trip to the tailor, another ten pounds lost, and the uniform was sagging again. When his weight finally leveled off, his alterations man must have been sad. Having multiple uniforms in several styles, George had been steadily ringing the tailor's cash register, but no more.

Standing watch as Reactor Technician was a breeze. The RT's station was Reactor Compartment Upper Level. However, with the reactor at power, exposure-limiting practices compelled the RT not to loiter in the vicinity of the reactor compartment. The amount of radiation he received during a six-hour watch was low. Nonetheless, it was the duty of every Navy man to minimize his radiation exposure.

The Reactor Tech went into Reactor Compartment Upper Level every half hour to monitor equipment and to log temperatures, pressures, tank levels, particulate radiation levels, and the like. When finished taking logs, he exited the compartment and spent his time in the Engine Room.

Just forward of Maneuvering sat an industrial two-warmer coffee maker. As RT, and a coffee drinker, I headed toward the coffee pot and passed George Short on the way.

"Want some coffee, George?" I asked.

"Sure," he said. "But it's old stuff. I can smell burnt all the way over here."

"I'll make some fresh," I said.

George snorted. "Don't make that weak shit you usually make." You see, George is my friend. OK, Chief, a brand new pot of coffee coming right up.

I filled the drip-through coffee filter to the top with fresh coffee grounds. That was more than double the standard amount of grounds. After the pot finished brewing, I filled a new filter with new grounds and ran the coffee through a second time. The twice brewed coffee had an amazing aroma; not quite coffee, not quite fertilizer, it definitely did not smell delicious. In George's porcelain coffee cup it was as black as outer space. An oil slick shone on its surface. I brought the cup to George because he is my friend.

"Just what you wanted, the same weak shit I always make."

George knew he had been swindled before the concoction reached his lips. Being an old salt, he drank the coffee as if nothing was out of the ordinary. After a sip, which he savored like he was evaluating fine wine, he remarked, "Good stuff." He drank the contents of the entire cup and asked for seconds. He took my little joke without blinking and pissed on my campfire at the same time. George Short was a man among men. George Short was also awake for the next thirty-two hours.

There are time-honored sagacities submarine sailors use to display a devil-may-care attitude toward authority. When confronted with the warning "you'll be in trouble if you do that," a submariner was apt to reply "what are they going to do, send me to sea on a submarine?" This was an ala carte smart-ass retort; any number of things could be substituted in as the horrible punishment.

What are they going to do, make me eat tofu? What are they going to do, make me shine my shoes? A favored expression in defiance of power was: What are they gonna do, take away my birthday?

In the Pacific Ocean, at a line passing north and south just east of Australia, lies the International Dateline. Crossing it while traveling east instantly turns Tuesday into Wednesday. When you cross it at midnight, June 12 becomes June 14 in an unimpressive moment. Everyone else in the world experienced a June 13th, you passed right through it. It's time travel for sailors.

I came into the shower and laundry space in the Stern Room to find George Short sitting on an upturned bucket. His elbows propped on his knees, his head in his hands, he looked dejected.

"George, you all right?" I asked.

"They finally did it. Eleven years in the Navy and they finally did it."

"Did what?"

"Took it away."

"Took what away?"

"My birthday."

Crossing the international date line had disappeared George's special day into never-never land. More than the loss of his birthday, George felt he had lost the game. "Sons-of-bitches finally got me."

Cruising into the Crew's Mess for dinner, I was happy to smell grilled steaks and baked potatoes. It was a favorite meal of mine and Bill Phillips was a great cook. When Bill was cooking a line formed back into the Control Room and curved around the periscope well, as hungry men waited for a seat to open up in the Crew's Mess.

When arriving during meal time to discover the absence of a line, I knew it was Grun on duty. His meals were bleak. It was hard to believe that the two cooks worked from the same recipe cards. A person would think one meal was prepared by Thomas Keller at the French Laundry while the other was fixed by a group of boy scouts with sticks over a campfire.

I loaded my plate with a T-bone, a couple of "trees"—broccoli stalks—and a baked potato with all the trimmings. Carrying perishable foodstuffs to feed a hundred men for an extended length of time required techniques that should not be used at home. Case in point: reconstituted sour cream. It looked a little off-color and its consistency was not exactly what I expected for sour cream, but my expectations were adaptable. As carpenters say: it looks good from my house.

I spooned a big dollop of reconstituted sour cream into the crevice of a split baked potato, right on top of a pool of melted butter. Oh, this was going to be good. I forked in a big bite of steaming potato with melted butter and enough sour cream with chives to make any comfort food epicurean swoon. Only it wasn't sour cream as I had thought. It was reconstituted horseradish.

My table mates were gearing up to perform the Heimlich maneuver on me when I finally got back my voice. I could only croak but it was enough to stop them from squeezing my solar plexus and blowing baked potato all over the crew's mess. The man sitting across from me had wisely vacated his position at the first sign of trouble.

The only up-side to mistaking reconstituted horseradish for reconstituted sour cream was that my sinuses had never been so clear.

At sea, news of the outside world came in snippets. The submarine slowed and came to periscope depth. We poked an antenna up out of the water and grabbed the compressed Associated Press wire transmission. The latest news was printed out on 8 ½ x 11 inch paper and left in the Crew's Mess for all to read. There were no in-depth articles, only headlines with a few clarifying sentences:

"THE CHICAGO BULLS ARE THE NEW NBA CHAMPS, BEATING THE BOSTON CELTICS 112 TO 108."

"MICROSOFT ANNOUNCES THAT INTERNET EXPLORER WILL COME AS A STANDARD PACKAGE ON ALL NEW PERSONAL COMPUTERS."

"EARLIER TODAY THE UNITED STATES BOMBED LIBYA."

"UNIVERSAL PICTURES ANNOUNCED—"

Wait a minute. Back up—what was that? We bombed Libya? Holy shit, are we at war? Why are we bombing another country? Questions flew hot and fast. The CO keyed the mic for the Ships Announcing System.

"This is the Captain. Many of you have expressed concern about an item in the news that stated that the United States has dropped bombs on Libya. I do not have details on that situation, but I can tell you that so far as I know, we are not at war." The microphone went quiet.

As a Special Projects submarine our job was already defined.

War or no war, not much would change for us. At least, not at first.

I imagined other fast attack submarines racing into the Mediterranean. Boats on their way home from long deployments turned around and headed back out. Upcoming shore leaves were canceled. We may not have been at war, but you don't wait for a Pearl Harbor repeat to get your chess pieces in place. If the Libyans decided to take on Uncle Sam with a frontal assault, they could be surprised by a hailstorm of missiles rising out of the sparkling blue Mediterranean Sea.

Those missiles would be launched by submariners who were really pissed off about being deployed three months longer than planned. I promise you, those missiles would hit spot-on and their targets would be DRT: Dead Right There.

During one deployment, Clem Davis and I held a private contest. The winner would be he who succeeded in causing the other to lose his cool. Simple childish taunting was beneath us, this game had rules. Forbidden was physical harm to the man or his possessions. Short-sheeting of beds, shaving cream under the pillow, that was frat-boy rookie stuff, and therefore disallowed. We aspired to a higher level of cruelty.

Bent on psychological breakdown, we tried to get inside the other man's mind and put a twist on it. The goal was to pressurize his patience until it burst at the joints. Breaking Jim was going to be tough. He seemed un-rattle-able. We'll see…

My ace in the hole was "the Sage." The Sage knew all about everything. No matter what the topic, the Sage would argue any point at great length. He would one-up any story. If I told of catching a 15 pound Bluefish on a fishing trip, the Sage would recall the time he caught a 60 pound yellow-fin tuna. If I fought the bluefish for 20 minutes, the Sage reminded me that it took him an hour to land the yellow-fin and the reel had been destroyed in the process.

The Sage's stories were so extraordinary and so varied that we dismissed them as baloney. There was no way for a man thirty years old, having spent the last twelve years in submarines, to have all the adventures the Sage claimed. We weren't calling the man a liar, only pointing out that an equation dividing his declared exploits by his time-in-life had to be multiplied by a negative truth factor to equal anything believable.

The mathematical probability of every story he told being one hundred percent accurate was roughly the same as twin brothers winning the lottery from two different states — on their birthday — two years in a row. Possible yes; but not likely.

My ploy to use the Sage against Clem was conceived when an electrical operator took ill. Quarantined in his bunk to aid his recovery and not infect the rest of us, he could not perform his watchstanding duties. Now shorthanded, the Sage volunteered to take double shifts. He stood watch as Engineering Watch Supervisor and six hours later took the sick man's watch as Electrical Operator.

Clem Davis was Throttleman for this watch section. He was now trapped in Maneuvering only a couple of feet away from the Sage for six hours at a stretch. Victory was so close I felt happy already.

After my watch as Reactor Operator, I did Reactor Controls Division chores. In the course of doing the maintenance I was assigned by the Preventive Maintenance Schedule, known in shorthand as PMS (an acronym which caused confusion at home in "what did you do at work today, Honey?" conversations), I dropped into Maneuvering to have a work procedure approved. Whenever things were going to happen out in the engineering spaces, Maneuvering needed to know about it. Standing by for several minutes while the EOOW looked over my paperwork gave me an opportunity to begin a conversation with the Sage.

"Last night I dreamed about a time I was surf fishing off Cape Cod. The dream was so real that when I woke up this morning I couldn't figure out where I was," I said.

"I have dreams like that all the time," said the Sage.

Clem was giving me the dirtiest look he could produce. He knew what was coming. I continued the thread.

"I remember like it was yesterday," I said. "I caught a 15 pound Bluefish that fought so hard my shoulders were sore for a week."

The Sage looked at his panel for a moment. Then slow and

quiet—as if remembering an event long past or perhaps inventing one—he declared: "I was fishing on a party boat out of Waialua, and caught a 60 pound yellow-fin tuna."

"Really?" I was astonished. "Tell me, tell me."

The Sage leaned into the tale with the air of a man just getting warmed up. My work procedure approved, I retreated from Maneuvering to do the weekly maintenance on the Nuclear Instruments. I grinned at Clem as he murdered me again and again with lightning bolts shooting out of his eyes. Having wound up the Sage, I could leave. Clem could not. My work here was done.

Maneuvering's door was open an hour later. As I soft-footed past I could hear the Sage laughing. He was enjoying his story, even if no one else was.

"When I finally got that fish on the boat, the skipper said he'd never seen anything so…" Clem confided later that the Sage told him the fishing reel was destroyed in the process of landing the fish.

I continued to wind up the Sage in Clem's presence for several weeks. I came close, but never broke Clem. He remained cool and calm for the entire mission. A vein on his forehead bulged a few times. The spoon he stirred his tea with bore bite marks, but the man never lost his temper.

Finally, a familiar announcement sounded over the ships announcing system: "Passing under the Golden Gate Bridge."

We were almost home. That marked Armistice Day for Clem and me. We called it a draw, shook hands, and never played that game again. I had almost gotten the better of Clem Davis, but not quite.

29

Rude Bastards

Rude: an otherwise gentlemanly sailor after six beers.
Bastard: the same sailor after eighteen beers.

— *The Submarine Sailors' Imaginary*
Book of All Knowledge

BEING A 1950'S era design, some eccentricities of seafaring tradition remained on *Seawolf*. Her engine room had two voice tubes; one between Engine Room Upper Level and Engine Room Lower Level, and one from Engine Room Upper Level to the Feed Station. An antique-sounding telephone bell announced a call. A funnel at each end of the tube, big enough to encompass the lower half of a face or the side of a head, isolated voice and ear from the noise of the surrounding machinery. Basically a two-inch pipe with a funnel at each end, it worked surprisingly well.

Ringing up the Feed Station and emptying a pitcher of ice water into the voice tube just as the man answered had been done many times. I tried it on Rob Kersch to no avail. Rob was too quick on the recoil. Only a rookie could be caught off-guard. After his first dousing, the lower level watchstander was wise to the game and for the rest of his career he would stand to the side of a speaking tube, not directly in the line of fire. Also, this man you have hit in the face with ice water will plot against you.

Practical jokes are most fun to play on your friends. The trouble with this philosophy was that your friends know where you live. To enjoy a long and happy life in submarines your practical joking must remain anonymous.

A Lieutenant Junior Grade was known as "the Huffer." This behind the back only nickname was given for his ponderous sighing. Apparently his mind was overloaded with weighty matters. On sitting, he would emit an exhausted sigh as if he were sitting down for the first time ever. Offered logs for review, he would accept them as a reluctant hero wearied under the burden of his responsibilities. He exhaled as if he were undergoing a chest examination.

"Deep breaths now," the doctor might say, placing the stethoscope on the lieutenant's chest. Air in—and—Huff! Out it came. Hence, the Huffer.

The Lieutenant looked reasonably fit but did not possess what anyone would call an athletic build. He was a little soft around the middle; his shirt did not hang in a straight line from chest to waist. The cloth peaked one button above his belt on its way south.

He became self-conscious about his weight only because the Nukes commented on it daily. The joke had started slowly, without intentional design. Men on watch for six hours at a time, performing tasks they are accustomed to and proficient at, sometimes occupy their extra brain waves with more entertaining thoughts. So began Operation Harpoon.

The mission: To find out how far the Huffer can be pushed toward the edge of his wits. The operation was known only to the Nukes in the Huffer's watch section, plus a few others. It was an uncomplicated plan.

One man or another said, "Sir, you better lay off the cheesecake," while pointing to the LTJG's middle.

The next evening another man said, "Roast beef for dinner tonight, Lieutenant. You really ought to stay away from the potatoes and gravy, on account of, well...you know, sir..." as he patted his own belly.

The Huffer protested at the innuendo. "I am not gaining weight!"

"Whatever you say, sir." Add an eye-roll and the point was scored.

The men continued to lay it on with a light touch, one small "observation" per day. For several weeks they kept it up. Propaganda is most effective when delivered as a constant, insistent drone. The Huffer began to accept the rumor.

The level of play stepped up a notch when enlisted Nukes enrolled an officer as an inside man. This lieutenant, code named Agent Orange, shared a stateroom with the Huffer. Before the Huffer climbed into his upper bunk, he hung his poopie suit on a hook so that the poopie suit hung directly in front of Agent Orange's bunk.

The poopie suit irritated Agent Orange each time he climbed out of the rack. He swung his feet to the deck and there it was; his roommate's blue poopie suit, right in his face. Worse yet, it was the crotch of the pants that were in his face. Men have a phobia about the crotches of other men's pants. Especially when they know those pants only get washed occasionally.

In opposite watch sections, the two lieutenants rarely saw each other awake. Agent Orange was reminded of the Huffer's existence only by this blue insult flapping in his face when he climbed out every morning.

When the half-asleep Agent Orange swatted the poopie suit away from his face, its brass belt buckle swung around and dinged him on the side of the head. Now he was irritated at the Huffer for hanging the coverall there, and mad at himself for not remembering that the belt buckle would bash him if he reacted roughly. He mentioned this source of angst to the reactor operator in his watch section as an explanation of the small cut just above his right eye.

This reactor operator knew of the scam being perpetrated on the Huffer. As a member of another watch section, it had been appointed to him to drop a weight hint on the few occasions when the RO encountered the Huffer. This independent third-party endorsement was designed to validate the Huffer's own watch section in regard to his supposed weight gain.

These guys were good.

Sensing a kindred spirit, by the end of the six hour watch the RO had enrolled the Lieutenant as a double agent. The LT was, of course, a member of O-gang. As such he was allied with the other side, the officer corps. He was on the same side of the coin as the Huffer. But now he was a useful tool of the enlisted men. He was Agent Orange.

Turning him was easy. No water-boarding, no jumper cables, no bright lights. He listened to the plot against his annoying stateroom mate and said, "I want in." So began Phase II of Operation Harpoon: Operation Payback.

A military web belt is built as a one size fits all arrangement. In manufacture a metal strip is folded over the free end of the belt webbing. This free end is looped around the waist, run through the buckle, and aligned to the buckle's edge with appropriate military smartness. The other end slides through an adjustable clamp on the back of the buckle to adjust the length of the belt.

Common practice was to trim the excess belt material to just behind the snap closure on the buckle's back. The loose end of a too-long belt wandering out from behind an otherwise neat appearance made one look as though his family tree might list only one last name.

Military men know many shortcuts to maintaining appearance. One such technique involves the web belt buckle. Aligning the metal end of the belt precisely with the edge of the buckle was tricky. As the belt was pulled tight in the buckle, the metal end retracted. To have the thing cinch up with the edge of the metal belt end aligned just so with the edge of the buckle required a try, then try again approach.

One solution was to fuss with it until the belt end and the buckle edge were aligned perfectly, and then leave them forever in place. One simply put on and took off his belt from the reverse side of the buckle. The Huffer subscribed to this procedure.

Each morning as Agent Orange rolled out of bed, while the Huffer slept, Orange carefully trimmed a quarter-inch length from the free end of the Huffer's belt. As the conspirators in the Huffer's watch section chided him over his imaginary weight gain, his belt was simultaneously being incrementally shortened.

"Jesus," thought the Huffer. "I am gaining weight…"

He began to eat more consciously, avoiding foods on Weight Watchers "don't touch it" list. No more pies and cakes made by the Night Baker. No cool, creamy soft-serve after a long, hot watch in the Engineering spaces.

An electrician reported observing The Huffer glancing through a book of healthy eating habits. The book had been laid out as bait.

The Huffer started a new routine of health. Sweating on the treadmill stowed in the Projects spaces was challenging. He tried to think of other things as he ran to nowhere. The treadmill was in an aft corner and out of the way. The Huffer ran his miles with a Walkman on his belt and oversized headphones over his ears. His eyes absently viewed the activity in front of him. During one of his runs he saw one sight he wished he hadn't.

A diver climbed out of a dripping wet suit. This Senior Chief diver was an old-school sailor with tattoos of naked ladies rendered in blue-green languishing on his forearms. The Senior Chief was also an old-school diver. He went commando under his wet suit; no skivvies. The Senior Chief peeled down to reveal a blue-green, six-inch diameter, seven-bladed propeller tattooed on each butt cheek.

When the conspiracy against the Huffer began, he carried a slight bulge around his middle. Now, a month later, the man was actually losing weight. At first it was only a couple of pounds, but the loss was becoming noticeable. As he became leaner he began to feel better about himself. But something bothered him about his lost weight. Just when he finally dropped a few pounds it seemed to come right back.

His measuring standard was his belt. Each day he tugged at it when he thought no one was looking. He was gaging its slack, or lack of. The scowl on his face revealed his consternation. He never knew about the secret agent man living below him. Agent Orange simply faded back into his previous life as a loyal member of the Officer Corps and as the faithful roommate of the unsuspecting victim. The worst part of being a great double agent is you can't take credit for a job well done.

Wild Bill the reactor operator stood his watch in Maneuvering with an electrical operator who had never been to sea before. The EO, Evan Weintraub, was a fresh faced kid. A nice guy, he believed everything he was told. For a submariner, that was a bad habit begging to be exploited.

Because of his open demeanor and friendly nature, Weintraub was a favorite target for jokes and other perversions of humor. He was trusting; he simply did not believe that anyone would deceive him, even in fun. It was our duty to educate him. Otherwise the poor lad would live his life continuously falling prey to unscrupulous salesmen sniffing out an easy mark. We were doing him a favor.

The Reactor Operator and the Electrical Operator "stood" watch seated next to each other at adjacent control panels in the Maneuvering Room at the aftermost end of the Engine Room.

On the occasion of Wild Bill being particularly quiet during the six hour watch, young Weintraub inquired if anything was wrong with the Reactor Operator.

"I haven't been laid in two weeks," grumbled the RO. Startled, the young Electrical Operator laughed.

"Two weeks! What are you talking about? We've been at sea two months!"

Wild Bill stared at the Electrical Operator until the man started to squirm. The Reactor Operator reached over and laid his hand on the other man's thigh. The Electrical Operator jumped out of his seat in front of the Electric Plant Control Panel as the full implication of the past few moments dawned on him.

"Electrical Operator, mind your panel!" growled the Engineering Officer of the Watch. Red-faced, Weintraub returned to his seat, suspiciously watching the Reactor Operator. He wasn't certain if the RO was kidding.

For the remainder of the mission, Wild Bill would come up behind Petty Officer Weintraub and lay a hand lightly on the man's shoulder. Upon noticing who the hand belonged to the young electrician would jerk out from under it.

"Don't touch me, you sicko!" He glowered at Wild Bill. Weintraub's cherubic face and easily flushed cheeks made the threat hilarious to bystanders. Unaware of the history behind the remark, they just laughed. They didn't need to ask details, they were patient. No one was going anywhere for the next month or more. With time on their hands it was more fun to wait and watch.

Nukes could be like a pack of hyenas on the Serengeti waiting for a "kill or be killed" battle to play itself out. Whoever the victor, the hyenas eat well.

Our language skills deteriorated at sea. Men not inclined toward foul language by nature or nurture slowly succumbed to the influence of poor company. Large vocabularies stored in the heads of smart men gradually shrank to a select few words. Swearing became sport.

Some men could use those few vulgar words to affect many different meanings. The Bluejackets' Manual will tell you that "Men are profane usually because they lack education..." Maybe, but maybe not. It took a smart man to communicate a thought completely by repeating the same words in different contexts. Ours was a game of intelligent discourse and wit rather than simple brutish profanity. It was an exercise of skill, like Van Gogh painting sunflowers with only one brush.

A certain Nuke machinist mate chief petty officer used the F-word numerous times in every sentence. The real F-word, not the ORSE F-word. For him"fuck" could be applied as any part of speech; noun, verb, adverb, or adjective. He often accomplished all these in the same sentence. My eighth grade grammar teacher—kindly, white-haired Mrs. Graham—would have required sedation after a conversation with the chief.

The chief's lack of decorum in his language use was so widely noticed that the entire ship's force would seek out ways to engage him in conversation while keeping score behind their backs. Men would brag on their scores.

"Yesterday I got 32 fucks out of MOO in about 30 seconds!"

"Damn, that's good. I got 58 out of him a couple of days ago, but it took five minutes." The acronym "MOO" spoken as one word, stood in for the chief's well-earned nickname: Master of the Obvious. The man was famous for statements only a mother should make.

When Brian McCoy smashed his thumb under Volume III of the Reactor Plant Manual, MOO said, "You should be more careful."

When Roger Brun spilled hot coffee in his lap and leapt to his feet cursing, MOO said, "Watch out, that coffee's hot." Thanks a million, Chief. Don't know how we could get along without you. MOOOC was his full name, Master of the Obvious Out of Control. We called him MOO for short.

Bitching became a pastime. A good sailor could bitch about anything. He'd even bitch about good things. Cursing as much as possible was one of the few viable vices onboard a submarine at sea. Bitching and cursing in tandem was an art form:

Jonesy, what's for lunch today?

"Those goddamn cooks roasted up some fucking seagull." (Game hens or chicken.)

What else?

"Biggest goddamn trees you ever saw and fuckin' gobs of mashed spuds."

"Trees" were broccoli. I've never seen in a grocery store a head of broccoli with six inches of bare stalk attached. This was the norm on the boat. Produce sellers with a Navy contract apparently were paid by the pound.

Any dessert, Jonesy?

"Pie, two kinds, dingleberry and this fuckin' chocolate shit that looked like something left in a diaper when the baby needs to go to the doctor."

Was it good?

"Fuckin'-A-right it was."

Cursing became a form of expression far beyond what more civilized circumstances would allow. Here vulgarity was normal. The coarsest talk you might hear in a civilian country club locker room would cause a submariner to laugh out loud at such an unskilled waste of the four-letter lexicon.

Hey Jonesy, hand me a coffee cup, will you?

"Fuck you, you fucking fuck. Do I look like your fucking butler?"

After the dinner dishes were cleared away in the Crew's Mess, the lights were dimmed and a movie was played. In the 80's this meant VHS video tape. Hollywood supported the troops with more than lip service by allowing the Navy to have films prior to their release to the public. U.S. Navy men at sea could watch the latest release at the same time it was in the theater back home.

"For U.S. Navy Shipboard Use Only," was the first screen shot. About once a year the *Navy Times* reported that a supply corps pinhead had been busted for bootlegging the Shipboard Use Only films.

Turns were taken on picking the movie from our library. To avoid the inevitable argument, the unwritten rules allowed the chooser to choose what he wanted. Those unhappy with his selection were free to leave and bitch about it elsewhere.

The movie chosen usually rotated between the latest action flick and the latest low-brow comedy. Robert Johns decided that in regard to our film selection, our couth needed tuning up. This sentiment brought on Classic Movie of the Week. Casablanca played six times straight so that every watch section and those doing maintenance after watch could view it. The movie was a revelation to the youngest men. New Bogie aficionados sprouted and grew.

"That guy is cool!" said a nineteen-year-old. "What's he done lately, you know, in color?"

We invented ways to segue the Epstein brothers' rapier wit into everyday speech. The dullest in our crowd simply used a line themselves. Instead of saying "thanks" when handed a manual or tool, they might say "Here's looking at you, kid." Or they might try to.

Overheard in the Torpedo Room: "Of all the joints in all the world, the gin had to come here…"

Stalled but not defeated, the man tried again."In all the world, in the joints that..." He was cut off by an exasperated voice behind a blue bunk curtain.

"Goddammit, Carlson, you are as fucked up as a soup sandwich. I swear, you would fuck up a wet dream. Just shut the hell up."

The highest level of play was putting out a line and giving the opening to the next man to fill in the blank. Caution was in order; there is no room in Naval Nuclear Power for informality.

"Request Permission to enter Maneuvering."

"State your business."

"To relieve the Reactor Operator..."

"Enter Maneuvering."

"...and to inspect your papers! Sir, I am shocked, shocked to find there is fission going on here!"

The on-watch Reactor Operator brought it back.

"Of all the Maneuverings, in all the submarines, in all the Navy, you walk into mine."

The Throttleman responded to the on-watch Reactor Operator.

"How extravagant you are, throwing away Reactor Operators. They may become scarce!"

The Engineering Officer of the Watch frowned. "Maintain the formality of Maneuvering," he ordered sternly. Then, *sotto voce*: "You despise me, don't you?"

The Maneuvering watchstanders instantly returned a unanimous "Yes!" A chance like that didn't come along often.

The originator of the banter looked at the EOOW with sincerity and delivered the line just as Rick did to Ugarte:

"If I gave you any thought I probably would—Sir."

The men returned to their business of turning over the watch in a formal and professional manner.

Next up was The *African Queen*. I could hardly wait for some choice Kate Hepburn zingers. She skewered poor Charlie Allnutt time and again.

30

Rank Privileges

The army here is in the best of spirits.
It's a good thing the men don't know everything.

— *Erwin Rommel, in a letter to his wife from Tunisia 1942*

Seawolf was heavy with rank; the most junior men on the boat were third class petty officers. In the surface fleet that rate provided a modicum of stature and privilege—a minuscule amount—but none on submarines and certainly not on a Projects boat. The security clearance required for submarines answering to the Office of Special Projects was too high for entry level positions.

Technical training for jobs in submarines was lengthy enough to ensure sufficient time in service for most enlisted men to rise to a petty officer rate before assignment to a nuclear submarine. My Nuclear Power School class graduation photograph, taken at least six months before any of those guys would arrive on a seagoing vessel, shows five second class petty officers, twenty-three third class petty officers, and one seaman. The E-3 must have screwed up along the way and lost a stripe.

The term "captain" is both a navy rank and an honorary title for anyone in command of a vessel. In the diesel boat era a submarine was thought of as a small command, warranting a commanding officer of lesser rank than a larger ship. WWII submarines carried sixty men and were often captained by Lieutenants. In the era of nuclear powered submarines the crew size doubled, but a Nuke boat was still a relatively small command when compared to an aircraft carrier numbering 6500 souls.

Cold War era submarine captains most often held the rank of commander. A submarine commander holding the higher rank of captain was uncommon but not unheard of. Even so, in 1985, Commanding Officer *Seawolf* was a Navy captain. Silver eagles adorned each collar point of his impeccably pressed khaki shirt, inspiring the colloquialism for his rank of "Full Bird."

Master chiefs, senior chiefs, and chiefs represent the top tier of the enlisted ranks. The chief's quarters was referred to as "the goat locker" by one and all. *Seawolf* rostered three times as many chief petty officers as a typical fleet submarine.

All this is to say that when time came to dole out an undesirable task, the bottom rung on the seniority ladder was still relatively high. Even the worst tasks went to those the Regular Navy might have considered to be above such work. Submarines carried no one low in seniority to do the grunt work. Everyone did grunt work. On submarines it was poor etiquette and worse, poor salesmanship, to utter the phrase "that's not my job."

Departments took turns in assigning one person to "mess crank" for each underway. A noun as well as a verb, a Mess Crank was a kitchen helper who did all manner of jobs, so long as they were mundane. For the most part this meant washing and cleaning. A Mess Crank was lost to his home department and belonged to the cooks for the duration of the loan.

In the Engineering Department this task usually fell to the least qualified person, partly because of his lack of seniority, and mostly because the department could little afford to lose a watchstander.

The senior Engineering Department enlisted man, the Bull Nuke, decided which individual's absence would hurt the least out of the four Engineering Department divisions. He made that recommendation to the unlucky man's Division Officer who made the assignment.

Nuclear war with the Soviets would seem a minor annoyance compared to the blow back from telling a petty officer second class that he was assigned to mess crank duty for three months. In the surface fleet that E-5 pay grade would have enough cachet to preclude such a lowly assignment, but on a Projects submarine a second-class carried no weight whatsoever.

When Lieutenant Reilly, the Reactor Controls Division Officer, chose Petty Officer Second Class Rick Traver as mess crank he knew it was not going to be a popular decision. Not with RC Division and certainly not with Traver. Traver had already served as mess crank previously which should have exempted him from the pool. Traver was qualified the same watchstations as every other member of RC Division, but unfortunately his sleeve carried E-5 chevrons while the other division members sported E-6 insignia. Those same "senior members" of RC Division would take up the slack created by Traver's absence on the watch bill.

Many of the E-6s never mess cranked, myself included. I showed up on the boat wearing First Class chevrons and was automatically exempted. Now Traver was suffering the same bummer twice. Sorry Rick, but better you than me. Such is life in the military.

To his credit, the Division Officer personally broke the news to Traver. He could have sent word down the chain of command, but some things are best done man to man.

Traver gave no credit for the gesture. If the betrayed look on Traver's face didn't give the Lieutenant pause, fear of retribution should have.

Petty Officer Traver did mess crank for three months, of course. Lieutenant Reilly found himself in an unnerving position.

In the way back, before Rickover walked the earth, a court flatterer named Damocles traded places with his king, Dionysus, to enjoy the imagined life he envied. Damocles's view on the joys of leadership lessened when he noticed a sharp sword hanging over his head by a single strand of horsehair. Dionysus had placed the sword to illustrate to Damocles the true nature of life as a decision maker.

So it was with Lieutenant Reilly. Each time the lieutenant sat for a meal he worried that his plate had been fouled by a vengeful second class petty officer in the kitchen. The lieutenant could imagine innumerable ways to accomplish the task leaving trace evidence only detectable through DNA analysis.

Traver amplified the effect by having the server repeat "Petty Officer Traver hopes you enjoy your meal, sir" each time food was placed before the lieutenant. Whether anything was done to the lieutenant's plate I do not know, but the possibilities stole the lieutenant's appetite. No one was gladder when Traver's kitchen time was up than Lieutenant Reilly. The young officer's khaki's were sagging from his weight loss.

Sharing grunt work only went so far. Faced with the choice of doing or delegating a task, someone eventually played the "Rank has its Privileges" card. If the receiver of an unpleasant task said, "Why me?" the giver had only to tug on his collar devices or point to his sleeve as if laying down a superior poker hand. The farther down the promotional ladder you were, the more unpleasantness was shoveled on your head.

While underway on the Mission of '86, Auxiliary Division—A-Gang—held a lottery to choose the person to go into Sanitary Tank #2 and investigate why the outboard valve was not opening. When operating correctly the outboard isolation valve allowed the contents of the tank to be blown overboard. The two heads in the Stern Room used by the Nukes flushed into SAN2. "Head" being Navy-talk for toilet, the direness of the situation should be clear. If we couldn't get rid of the waste in the tank, we could not continue to fill the tank. The aft heads were immediately closed to new business.

A chief suggested placing a block of cheese on every table on the mess decks. Eating mass quantities of cheese was calculated to buy time to sort out the sanitary tank malfunction. While Reuben Goldberg himself would have admired the direction of the Chief's thinking, a short term stopping up of the crew was not solution enough. We had a long way to travel and the ugly job of fixing the problem was the least bad choice.

Someone would have to put eyeballs and hands on the valve to see what was what. The unpleasantness of the troubleshooting was exacerbated by the location of the valve. The remotely-operated valve was located inside the tank. A-gang men would have been happier to lay unfriendly hands on the engineer who designed the system, and send him into the tank.

Asking for a firing squad test target would have produced more volunteers. Every man in the division was equally qualified for the job and equally against his doing it.

There exists a conflict of memories as to who pulled the short straw. I do know that to mistakenly credit the wrong man with going into the sanitary tank would be a crime against the man who did go into a full tank of raw sewage. As of the date this manuscript was put to bed, I had yet to connect with the man reputed to have gone into the tank. I wanted to verify the whos and whats before naming names. Perhaps that wrong can be righted in the future. Even though one man went in, the whole of Auxiliary Division was involved. Up to their eyeballs, you could say.

At any rate, a man, a human being, a mere mortal, went into that chest-deep witch's brew of black death. And—he came out alive.

The blockage cleared, SAN2 was air-pressurized greater than sea pressure and emptied to sea as designed.

A short time later, Auxiliary Division Leading First Class Dave Lyon marched into Maneuvering. The reactor operator on watch, Petty Officer Second Class Mike Plant, had misplaced his wallet several days earlier and was depressed about not finding it.

Plant irritated his fellow watchstanders by mentally retracing his steps out loud hoping for a revelation as to the wallet's whereabouts. The very definition of a closed environment, the number of places on the boat he could misplace anything was limited. He recalled looking at a photograph kept in the wallet while he was seated in the aft head. He was certain he had placed the wallet in his poopie suit back pocket, finished his business, flushed, and that's where the trail went cold.

Being off-watch, Plant had slept through the SAN2 drama and was unaware of the tank diving adventure.

"I believe this is yours," Dave said coldly, extending the sodden wallet that had blocked Sanitary Tank Number 2 Outboard Valve. The blue nylon wallet had undergone a thorough washing and smelled sharply of disinfectant. Surprised by his good fortune, Plant exclaimed, "Hey, thanks! Where did you find it?" He ripped open the Velcro closure and a look of confusion came over his face. Plant tentatively sniffed at the wallet and recoiled angrily.

"Goddammit, Dave, it smells like Clorox! You asshole, what did you do to my wallet?"

Dave Lyon glared at the Reactor Operator for a long moment. His lips moved but no words came out. Veins on his neck pulsed; he brandished a menacing finger at Plant. Still no words came. He shook his clenched fist in Plant's face. Giving up on speech, he spun on his heel and left Maneuvering.

Engine room watchstanders stood at attention and solemnly saluted as he passed. "Fucking studs, those A-gangers," muttered the Engine Room Supervisor as he rendered slow-handed applause.

Without a backward glance, Dave strode down the passageway with dignity, his head high, waving his middle finger to the Engine Room and the world entire.

31

Halfway Night

The Chief says the world can't end today because in Moscow it's already tomorrow.

— *Submarine Sailors' Book of All Knowledge*

OUR MISSION WAS defined by accomplishment rather than time. The clock meant nothing and the calendar was only useful as a conversation piece.

Engineering Laboratory Technician Jim Stanton had an idea. "We should get home by the middle of the summer. It'll be warm. How about a bunch of us go fishing out of Half Moon Bay when we get back?" For the rest of the mission Jim planned the trip. Selective about fishing companions he kept the expedition secret. There were no dates involved in the plan, not yet. Our return date, yet unknown, was the day our lives would resume.

For some, life would happily take up where it left off. Some men would be greeted by good news, new babies, and warm homecomings. Others would return to divorce, death in the family, wrecked cars, and various domestic catastrophes.

Until then, we kept waking up, working, and going to sleep. We did plan in advance for one thing: Halfway Night. Halfway meant the mission was halfway completed. Approximately of course; round numbers, a ballpark figure. It was close enough for us.

Three days notice was given to the crew that Saturday night was declared Halfway Night. Whatever the state of morale at the time of the announcement it was higher afterward. The celebration was nice, but the significance of "halfway" made people happy. The back half of the mission gave us a long way still to go, but it was progress. We took what we could get.

Planning started for the festivities. Every division worked on a surprise skit or joke.

In the Control Room without urgent business, I stuck my head into Radio. The radiomen's shack was squeezed into a corner of Control. The space housed the submarine's radio gear and the on-duty radioman. It was the size of a broom closet wedged under the stairs.

"Hey fellas, what's happening?" I said, as on previous occasions.

I expected to hear "Hey, come on in, how's it goin'?"

Instead, I heard "Shhh!" This was directed not at me, but at others crammed into the space. I was offered as noncommittal a reply as was possible.

"Things are fine, here." Blank faces stared at me, offered nothing. They were the faces of men interrupted while planning secrets. Okay, gents, I'll be leaving now. Carry on.

For the next week, similar clandestine meetings were held all over the boat; planning, plotting, snickering, twinkling-eyed meetings. Walking up on men deep in conversation caused them to go quiet. They moved away or glared at me, sending the vibe "Don't you have someplace else to be?"

Finally the big night arrived. Dinner was exceptional. Each table in the Crew's Mess sported platters of T-bone steaks, and King crab legs sixteen inches long. It was all you could eat. Every few minutes another plate of steaks arrived at the table. Tom Johnston, who had a waist like an Olympic swimmer, ate three 16-ounce T-bones. Crab legs were replenished as well. Anyone who didn't leave the table stuffed to the gills just wasn't trying.

A special cake had been made in the shape of the submarine; the captain cut it to cheers and applause. Everyone got a piece.

The watch was relieved by men who waddled with the sated, happy stupor of victorious knights returning from the king's feast. The only thing missing was wine and wenches. Once dinner had been cleared the entertainment began.

While at sea, grooming requirements were relaxed. Who the hell was going to see us? Many let their beards grow. Some grew wild; some were carefully trimmed and curried. Some looked like mangy dogs half scalded.

The honor of "Best Beard" was given to the handsomest beard. "Bushiest Beard," also was awarded. The "Scraggliest Beard," was honored as well. Submariners are equal opportunity insulters. Everybody gets a share.

The Goat Locker put on a skit comically mimicking O-gang. We all laughed, officers included, at the joke and at the chiefs putting on the show. Most of the officers recognized their personal quirks and characteristics on display. If not laughing at their own foibles, they enjoyed seeing their colleagues satirized.

Some blueshirts put on a skit comically mimicking the chiefs. We all laughed at the joke, chiefs included.

A couple of acoustic guitars came on the trip with us and were played to accompany original songs. The songwriters in the crew were balladeers, singing crusty ditties about things that had gone wrong thus far in the mission. Tonight it was easy to laugh at what had seemed so serious only a few days before. Troubles are only serious until you know the outcome. Once a favorable resolution is in, the previous drama becomes public domain for humor.

After the music came a Pythonesque reading from the "Book of Armaments." The joke was on Charles Kane who, during a drill, had trouble understanding the word "three" over the sound-powered phones. He had finally understood the communication with only a modest amount of shouting. Since then, the number three had been a joke in itself. It was never spoken as "three," but as "THREE!"

Everywhere Charlie had gone since, catcalls of "THREE!" followed him. Charlie did the reading. It went something like this:

"And the Lord spake, saying, "First shalt thou take out the Holy Pin. Then, shalt thou count to THREE! No more. No less. THREE! shalt be the number thou shalt count, and the number of the counting shall be THREE! Four shalt thou not count, neither count thou two, excepting that thou then proceed to THREE! Five is right out. Once at the number THREE! being the third number to be reached, then, lobbest thou thy Holy Hand Grenade of Antioch toward thy foe, who, being naughty in my sight, shall snuff it."

We howled both at the reading and for our love of the movie it was plagiarized from. How can you not like a guy who publicly ridiculed his own mistakes?

Did I mention wenches? Our feasting and revelry had no wine, true, but we did have wenches. Sort of, in a manner of speaking, leaving room for imagination. A Grand Canyon sized room of imagination.

Think of Elizabethan Era theater. Think summer camp. Thinking of *La Cage aux Folles* would be giving our production values stupendously too much credit.

A beauty contest was held for men dressed as women. I know, I know… I'm in lead shoes on thin ice in recounting this event. Lest the world think submarines were indeed, as skimmers joke, an underwater gay swingers club; let me simply say this—No.

But now, on Halfway Night, parading through the Crew's Mess one by one was what all the secret sniggering had been about. Each division conjured an entrant, whether volunteered or conscripted made no matter. Anticipating this evening experienced men had smuggled onboard bits of clothing and lingerie borrowed or stolen from wives and girlfriends before leaving Mare Island.

Some divisions tried to corner the market on humor and put forth their homeliest man, complete with mustache, dressed in lingerie. The award of the "Ugliest Woman" went to an Auxiliary

Division machinist mate with body hair to match any gorilla. At this vision the audience screamed in horror and made as if to gouge out their eyes. Let us speak no more about that.

Then M-DIV's contestant entered through the watertight door. I refrain from using his name on the off-chance that he is now a respected member of his community. We'll call him "Stephani." Stephani was early twenties, very early twenties, slender, with a smooth complexion. He stepped through the watertight door into the Crew's Mess and the laughter valve slammed shut.

In addition to the sea foam green teddy, garter, and stockings, someone's wife had donated makeup. It was inexpertly applied but the effect was nonetheless startling. My brain stopped working as I tried to figure out where this woman had come from. I knew better, I knew it was one of us; but my brain had stalled. The other contestants were caricatures, who they were was obvious and made the joke funnier. But who was this woman—man—person? It took a full minute of staring before I figured it out. He looked good, really good.

Let us speak no more about that.

"Stephani" won Best Woman undisputed. I doubt that award is on his "I love me" wall at home. But then again, who am I to judge? Maybe he's more broadminded than I presume. For all I know he may still have the high heels. I hope he looks as good as he did then…

32

Technical Matters

For want of a nail the shoe was lost.
For want of the shoe the horse was lost.
For want of the horse the rider was lost.
For want of the rider the battle was lost.
For want of the battle the kingdom was lost.
All for the want of a horseshoe nail.

— *Traditional*

ONE OF MY current drinking buddies is a retired Army infantry officer whose first posting was 1969 Vietnam. Wayne's stories of life in the field always validate my choice in going Navy. During my military service, I slept on clean sheets over a decent mattress and never had clothing holed by a bullet. My Army buddy cannot make the same claim.

An army, so Wayne tells me, suffers from bad weather, rough terrain, the jungle, and the limitations of its supply system. In

Vietnam an infantryman spent a lot of time knee-deep in mud and water, as well as being regularly soaked with rain and poached to perfection in sweltering heat.

Now, in 2012, an infantryman in Iraq or Afghanistan endures sand invading his most personal spaces. Instead of jungle humidity, he roasts in "a dry heat." Armies achieve objectives by pushing through these irritations as well as the ever shifting situation of a mobile enemy shooting at them in a very personal way.

A submariner did not suffer the same type of physical hardships. When we deployed we took our warm bunks, hot showers, and a kitchen with four cooks with us. A night baker kept us "knee-deep" in fresh pies and cakes. Ice cream machines dispensed soft-serve on demand.

The physical challenges that most irritated submariners were a result of over-eating and under-exercising. Food was served four times each day on the boat. I came home from my first extended deployment to find that while I was away the Levi's I'd left at home had shrunken in the waist.

"Honey, what happened to my Levi's? Did you wash them in too hot water?" I asked my wife.

"No," she said. "They've been in that drawer since you left." Uh-oh, that's bad news.

My great physical challenge was to not eat my fill every time food was on the table. I learned to eat one meal a day and settle for a snack at other times.

When my friend Wayne threw his first hand grenade in Vietnam his hands were sloppy wet from the rice paddy he was standing in. He pulled the pin and the grenade slipped out of his hand disappearing into knee-deep muddy water somewhere near his feet. He managed to find the grenade and get rid of it in the right direction a moment before it exploded. He got better at hand grenades.

My main tribulation was too much pie and ice cream. Army hardships and Navy hardships are worlds apart.

A submarine is all about machinery. Modern sailors are more engineer geeks than able bodied seamen. The average enlisted sailor knew far more about electronics, hydraulics, thermodynamics, and radio waves than he did of ocean currents and celestial navigation. The crew's ability to achieve its objective had everything to do with the machinery. Machinery controlled the ship, machinery provided a life supporting environment, and machinery gathered the intelligence data we sought.

Should we be attacked, machinery would be employed to evade and destroy the enemy. In that instance, if the machinery failed, or we failed to use it well, we would be sunk (i.e.—we die).

Machinery also made our life relatively pleasant in comparison to our infantrymen cousins trying to sleep in the rain. In addition to the galley—nautical talk for the kitchen—we had heating, cooling, and hot and cold running water. We had flushing toilets, just like real people. No burning of shit or squatting in the bush for us. There was not a shovel to be found on the submarine.

When a problem occurred on the boat, it was usually related to machinery being disrupted. An argument can be made that a hole blown in the side of the boat is essentially a machinery problem. The explosion would have shut down or disabled machinery needed for control of the ship. Water coming into the boat was a technical issue of added weight and rapidly diminishing buoyancy. Machinery would be used to remove the water, hopefully removing it faster than it was coming in. If not, we sink (i.e.—we die).

It was simple physics, it was mathematical. Keeping the machinery running was everything.

As you will recall from our earlier conversations, any unplanned event that caused all or part of a system to shut down was termed a casualty. In an army one thinks of a casualty as a soldier taken out of action by injury or death. In the context of a naval vessel, the casualty was the affected equipment rather than a person.

A mathematician would be required to calculate the number of possible engineering plant and general ship casualties. Each system had likely faults that could be anticipated, for example a loss of electrical power. There was also the chance of malfunctions that engineers in swivel chairs had not foreseen.

Not only was there a large number of possible casualties, but each of those possibilities could vary in severity. The difference between a leak and flooding is titanic, pun intended. Immediate actions required to recover from those different casualties—different sizes of a similar problem—also varied in severity. Erring conservatively was desirable, yet no one wanted to cry "wolf" unnecessarily.

The Machinist's Mate Jerry Visak manning the Feed Station in Engine Room Lower Level phoned into Maneuvering. "Maneuvering, Feed Station, I have water coming into Engine Room Lower Level." Apprehensive to call away "flooding" prematurely—flooding being a major call in submarines—the Engineering Officer of the Watch bypassed the Throttleman phone talker and grabbed his own sound-powered phone handset.

"How much water? Is it a leak or flooding?" The communication that came back from the Feed Station is only partially reproducible in this civilized text.

I paraphrase: "There's a shitload of fucking water coming in down here!" The Engineering Officer of the Watch immediately called away "Flooding." You don't fool around with water coming into the people tank. It's all bad, especially a shitload. Too much water in the boat and we sink (i.e., we die).

Each casualty had written procedures. Immediate actions for each possible casualty were committed to memory for each watchstation before a man was deemed qualified to stand that watch. When the casualty was announced by the Officer of the Deck in the Control or the Engineering Officer of the Watch in the Maneuvering, every watchstation in the entire boat carried out his immediate actions autonomously.

Once the situation was stabilized, the watchstander pulled out the proper reference manual and verified by the book that all his immediate actions were complete. He then performed the listed supplemental actions, designed to prepare for recovery of the lost functions.

In real life one fault often caused a failure elsewhere. This was especially true on an old boat like *Seawolf*. Even equipment working properly and tested regularly had seen years of use. An electrical transient could put an abnormal strain on electrical circuits, so that instead of being absorbed, the added strain caused another component to give up the game and so begin a new cycle of troubles. This was known as a Compound Casualty.

A submarine crew trained for all scenarios that have happened to a boat, could happen to a boat, or at least been thought of as possible even if unlikely. Even so, there are compound casualties that become massively tangled with what should be done first, second, and third. This step couldn't happen until that step in the other procedure was done, etc, etc.

Cool and knowledgeable heads were required to sort it out. In these cases the Reactor Plant Manual, the inspired holy word of Hyman G. Rickover, advised: Focus on the big picture and do the best you can. As an example consider the following scenario.

A leak in 400-psi hydraulics shuts down the hydraulic plant, causing each load powered by 400-pound hydraulics to go limp. These loads include the control surfaces of the submarine; the diving planes and the rudder. Besides "diving," the diving planes also can be pointed toward "rise." Yes, that might be useful in the very near future.

The mist of hydraulic oil blasted into the atmosphere is a fire hazard. Fire is always bad on a submarine. Fire is a monster eating up the same air we need to live. A hydraulic oil fire is a particularly serious problem. The hydraulic oil is under pressure and atomizes in the open air becoming an explosive concoction of petroleum and oxygen, heavier than but similar to the fuel-air mixture used by your car engine.

There is also a respiratory issue for the soft-tissue humanoids in the space. The air is contaminated. We strap on Emergency Air Breathing masks, EABs, and hook up to clean, piped-in breathing air.

If the loss of 400-pound hydraulics caused a diving plane to jam in the Full Dive position while we were answering a Full bell, the ship will head toward the bottom of the sea at full speed. This is a problem screaming for attention. We must quickly deal with our ship control troubles before they take the submarine to an unrecoverable depth (i.e. ,we die).

At the same time we could be fighting a hydraulic oil fire while wearing constricting breathing apparatus that inhibit movement, visibility, and communications. The best firefighting method is to prevent the fire from starting. Isolate the hydraulic leak and ventilate the space ASAP.

A situation like this can get out of hand if everyone is not on their game. It could easily become the real-life incarnation of the parable of the horseshoe nail. Putting out the fire in a timely fashion (i.e. - we don't die by fire) would be an unremarkable achievement if the boat goes to the bottom (i.e. – we die by drowning or are squashed into oblivion by sea pressure).

Anticipating scenarios like this kept us sharp. This is a prime example of why the submarine force requires smart people making good decisions quickly. That is the result of arduous training regimens and constant dedication to improvement.

Touring WWII submarines I was amazed to see the same interior communications equipment I knew from the 1980s. I assume the newest submarines use the same gear. Sound-powered phones were sturdy and required no electricity to operate. That's a clear advantage in emergency situations when the electricity supply is lost.

Sound-powered phones transmitted human voice with a tinny, robotic tone that took some getting used to in order to understand what was being said. Nuance was difficult to distinguish. Similar sounding words with different meanings were avoided. Superfluous words were not used.

Under stress, the human voice tends to rise in pitch. A low-pitched voice was clearer over the phones. An excited voice screeching into the phones certainly communicated concern, but communicated facts poorly.

"Holy jumping cow! There's steam everywhere! I can't see! Shut the valve! Shut the valve!" conveys urgency, but not information. Clarity—and speed—were better accomplished by a clear voice saying, "Steam leak, Engine Room Upper Level, starboard side." The Reactor Operator, needing no further direction, would remotely shut valve Main Steam One. All watchstanders would take appropriate actions.

You don't have to be a linguistics expert to imagine the trouble that could follow by mistaking "increase" for "decrease." In place of those words "raise" and "lower" were better choices. Submariners shunned the word "close" and "closed," preferring the more definitive sound of "shut."

Brevity and right-to-the-point statements were the soul of the sound-powered phones. Instead of, "I didn't understand you; please repeat what you just said," the request would go back to the sender, "Say again." Perhaps with a specific suggestion for transmission improvement, as in "say again louder," or "say again slower."

On one occasion of a high-pitched, indiscernible communication over the sound-powered phones it was Mike McGauley who responded with the pithy instruction: "say again more manly."

Generating electricity was a full time job on a submarine. High pressure steam spun turbines bolted to generators that produced electricity at 450 volts AC 3-phase. This power was broken down into lower single-phase alternating current voltages as well as being converted to direct current in various voltages.

Seawolf was built with seven AC generating machines which could be connected in parallel to supply one electrical grid. Normally, not all the generating machinery was connected together; the electric plant was "split" for reliability. This kept the whole boat from going dark on account of one failed machine, component, or load.

In your house, a simple wall switch connects the bathroom light bulb to your home's electrical grid and the bulb illuminates. Connecting the higher voltages and amperages used on a submarine required more sophistication in equipment and method.

Generators produce a rotating phase component that must be matched by all the generating machines operating on the same grid. The phase of the oncoming generator must be synchronized with the rest of the electrical grid prior to Shorting the connecting breaker. The electrical grid could not operate with one generator out of phase.

The breaker was designed to disallow that situation. If the Electrical Operator attempted to bring a generator on line while its output phases were out of sync, that generator's output breaker would trip open automatically before it could shut. (The term "fail safe" should come to mind.)

When synchronized properly, a generating machine brought on line took on electrical load smoothly. In Reactor Compartment Upper Level, I was made aware of an Electric Plant shift by the Engineering Officer of the Watch's 2MC announcement: "Shifting the Electric Plant."

Even though its operating switch was in Maneuvering, a heavy clunk in Reactor Compartment Upper Level informed me that the remotely controlled breaker had indeed shut. The lights didn't flicker; there was no disruption to the electrical status quo. The Engineering Officer of the Watch announced the new electric plant lineup.

"The Electric Plant is in a Full Power Lineup on Number One and Number Two Ships Service Turbine Generators." Engineering watchstanders monitored their equipment for the expected changes.

Submarines carry a lead-acid storage battery that is carefully maintained and kept charged. Your car battery is of the same type although in miniature. The submarine Main Battery was a large battery, 126 cells, each cell five feet tall and weighing close to a ton. The battery lived under the crew's mess with access through a hatch in the deck between two dining tables. It delivered 250 Volts DC at enough amps to run the equipment required to start the reactor. When electrical generating machinery went down, the battery kicked in automatically. It provided DC current to motor-generator sets which converted the stored direct current into alternating current used by the reactor plant machinery.

A motor-generator set is an ingenious device. By definition, a motor converts electrical energy into kinetic energy, or more practically speaking, into motion. Conversely, a generator converts motion into electrical energy. A motor-generator is an Alternating Current machine mounted on the same shaft as a Direct Current machine. Put DC in one end and get AC out the other. The DC end acts as a motor and drives the AC end as a generator. It also works in reverse. Put Alternating Current in, the AC end acts as the motor, the DC end as the generator; out comes Direct Current.

If AC power should drop off for any reason, the motor-generator continued to spin uninterrupted in the same direction. The flow of power shifted direction, and the machine switched to generating AC from the DC powering the machine.

If this is the first time you've heard about motor-generators, this AC/DC talk can be confusing. Just remember:

AC IN - DC OUT, OR

DC IN - AC OUT.

ONE OR THE OTHER, ALL THE TIME.

Seawolf had three Ships Service motor-generator sets. Number One and Number Two were located in Engine Room Upper Level; Number Three in Reactor Compartment Upper Level. She carried two Ship's Service Turbine Generators and two Coolant Turbine Generators. That made seven machines capable of generating alternating current. Not all had to be used at the same time, but enough machines must be run to give us the kilowatts we needed.

The electrical generating machines belonged to Electrical Division, or E-DIV, the electrical operators, also known as EOs, Nuke electricians, or simply as electricians. They maintained the electrical machines, and controlled them from the Electric Plant Control Panel—the EPCP—in Maneuvering. If the lights were to go out, these were the guys we counted on to fix it.

33

Lights Out

A ship in harbor is safe
but that is not what ships are built for.

— *John A. Shedd, Salt from My Attic*

GETTING THERE WAS the job of the whole crew, Engineering department and Nose-Coners. The Nukes pushed the boat and the Coners steered it. Now that the submarine was on site, the Projects Weenies came to center stage.

The work done by the Projects guys was unknown to Nukes. During operations, the watertight door at the forward end of the crew's mess leading to the Projects compartment was shut. A sign read, "Conducting Projects Operations. Stay Out."

The Projects space was located between the Crew's Mess and the Torpedo Room. The Torpedo Room was the forward-most compartment in the boat and the largest enlisted berthing space.

Before beginning a Projects run, the Officer of the Deck passed a warning over the 1MC:

PROJECTS OPERATIONS WILL BEGIN IN 30 MINUTES. ACCESS THROUGH THE PROJECTS COMPARTMENT WILL BE SECURED.

Men adjusted their stride to get to whichever end of the blockage they wanted to remain in for the next several hours.

The Nukes of *Seawolf* gladly fell into an easy routine. Most of us slept aft in the stern of the ship. Our living space and workspace, as well as the crew's mess, were all aft of Projects. The periodic inability to pass through Projects to the Torpedo Room was not an inconvenience to us. The Projects Department could beam itself up *en masse* to the starship Enterprise and the Nukes would never know or care about it.

Every six hours a fresh batch of Nukes rolled out of the rack, ate whatever meal was offered—appropriate to the 24-hour clock—and manned their watchstations. After watch they performed divisional responsibilities per the Preventive Maintenance Schedule.

Each man attended to his assigned collateral duties. Even a man fully qualified for his rating always had advanced quals to pursue. Studying for advancement to the next rate was required. Professional advancement was a military duty; lack of progress was a condition both unacceptable and punishable.

A ten or twelve hour workday put in, a man might eat again and spend an hour or two in relaxation before rolling back into his bunk. Thus passed days and weeks; unremarkable except for new Xs added to the calendar.

Several weeks into this lazy routine, Electronic Technician First Class Chris Berry awoke with a start. Unsure, he lay quiet. Something was not right. He swept open the blue curtain around his rack. It was dark in the berthing space, and quiet.

Abnormally dark and abnormally quiet.

John Uran heard the metal hangers slide in their aluminum track as Berry's curtain opened.

"You'll need this," Uran said, handing a flashlight into the dark cavern of the bunk. Switching on the light, Berry saw the flashlight was painted red.

"I can't use this!" he protested. "It's an emergency flashlight!" Red flashlights were stationed throughout the submarine. Using one in a non-emergency was only slightly short of a capital crime.

Uran's reply was grim. "Trust me, you can use it."

Two minutes before had been business as usual in the engineering spaces. One of the two ship's service turbine generators was down for routine maintenance. E-DIV men covered in carbon dust were shoulder-deep into the de-energized generator's access ports.

The brightly lit engineering spaces were filled with the sounds of machinery humming and whirring and pumping as it should. The engineering watch standers were about their business, monitoring, adjusting, taking logs, and sipping black coffee that had been too long on the warmer.

The lights dimmed without warning. The steady-state rhythms of the machinery were interrupted by groaning and growling of that machinery under duress. Every piece of running equipment made fierce noises, as if it were on the verge of seizing or blowing apart.

Horrified, the watchstander's heads snapped left and right, listening and sensing; scanning for data to explain what the hell was happening. The lights dimmed again and blinked out.

In the same moment, the machinery noises dropped in pitch as rotating machines were de-energized. Unpowered fans and motors wound down in revolutions on their way to a complete stop.

The steam turbines driving the main engines, the on-line ship's service turbine generator, and the two coolant generators, were now spooling down, deprived of steam and relieved of their loads. The propeller shafts stopped turning.

The main coolant pumps were in Reactor Compartment Lower Level which was sealed shut during reactor operation. Isolated as they were, they could not be heard, but those big pumps were also spinning down to a stop.

The reactor alarm siren was normally unnerving loud. It was designed to wake the dead and gain their full attention, a job it usually accomplished magnificently. Now, without sufficient voltage, rather than a full volume wail, the alarm siren rendered a pathetic growl that never reached its full intensity before I switched it off.

"Reactor Scram!" I called out. The Reactor Scram light on the Reactor Plant Control Panel was barely lit. Two tiny light bulbs behind the red indicator were glowing feebly. Reactor power should have been going down steadily, passing through the Intermediate Range into the Source Range. Instead all power ranges displayed zero. I had never seen nor heard of a reactor behaving so. My indications were confusing. Actually, it was the lack of indications that was confusing. Still, everything pointed toward shutdown. When submarine systems failed, they failed in the safe direction. Strange indications aside, the reactor was definitely shutdown.

"Main Coolant Pumps Two and Four indicate off!" I called out as the pump's indicating lights blinked off. "Shorting Main Steam One, Shorting Main Steam Two." Turning the remote switches to shut the air-operated steam stops that were in Reactor Compartment Upper Level, I reported:

"Main Steam One indicates shut! Main Steam Two indicates shut! "

They did but barely. Their indicator lights glowed faintly. The meters on the RPCP were dead. They all indicated their de-energized reading. Some were pegged low, some pegged high, and some pointed to mid position. Warning horns that should have sounded like a loud car horn instead gave only a low moan. Every instrument on the panel had failed.

Jeff Parkin, the Electrical Operator at the Electric Plant Control Panel beside me was also busy. When the initial electrical surge occurred he leapt out of his seat, his hands poised above the desk section of his panel. His eyes swept over the current and voltage meters. All seven generating machines were represented. The meters were all swinging wildly, except the ones for one off-line generator under maintenance.

"What the hell?" the EO had exclaimed as the compartment lights fluctuated between bright and dim like a Halloween joke.

"Sir, I'm not sure what's happening here, the whole electric plant is going crazy, it has to be the TG, it should have tripped off on its own, I gotta trip it off, that should put the plant in a safe condition, yeah, it's gotta be the TG!"

Having processed the bizarre information as best he could, he came to a decision. "Sir! Opening the breaker for Number Two Turbine Generator." And so he did.

Opening the breaker and disconnecting the ships service turbine generator from the electrical buss was the right move. Unfortunately, the expected positive change to our electrical circumstances did not occur. Instead, the EO called out, "Number Two Motor Generator just quit!" That was when the lights went out altogether. Lights in the entire submarine went out in a flash of blackness.

For the briefest moment nobody moved. In the darkness a voice rife with disgust spoke what everyone was thinking, "Shit..." Then people got busy.

Later, in the debriefing and lessons learned stages of the casualty, it would be hard to determine in which order things had occurred. The beginnings of the event seemed to happen all at once, and all within ten seconds time.

The inside of the submarine, normally lighted bright as day in work areas with a murky red glow in sleeping quarters, now demonstrated darkness of a density known only to the blind and to coal miners. Flashlights pierced the dark, throwing exclamation points toward gauges, meters, breakers, fuses, and everything else that would tell a story. Darkness wrapped around me like a hot blanket.

The guys who wore mini flashlights on their belt next to their Leatherman tool holster I had considered geeks until a minute ago. Now they seemed like really smart guys.

The silence of the machinery at a standstill stood out in bold relief against the backdrop of urgent voices that filled the engine room and Maneuvering. It was a scene straight out of *Das Boot*. Alarms, reactor sirens, and urgent voices were familiar sounds, but all the machinery shut down at sea was unsettling. That was very bad news.

Announcing systems had failed due to lack of power but we still had the sound-powered phones. The Engineer Officer was in Maneuvering within two minutes, the Executive Officer not far behind.

A reactor scram, while not the preferred way to shut down the reactor, was by no means the end of the world. All reactor operators had scrammed manually, been scrammed automatically, and recovered the reactor plant until the process was embedded in their being.

All the machinery in the boat was not supposed to stop on account of a reactor scram. Redundancies were piled on top of redundancies to prevent precisely that condition. The electric plant was run split to protect a vital buss. Only non-vital loads should have dropped away, as was their design and definition. The lights should have stayed on. When the lights did go off, battery operated emergency lighting should have energized. This present mess was something new. A fresh hell.

Men hustled to work out what had happened and what to do about it. Flashlight beams whipped through the dark like a laser show as breakers, fuses, gauges, meters, valves, and temperatures were interrogated.

The starboard ship's service turbine generator had been down for maintenance. The electrical operators were working frantically to reassemble the machine.

Information poured into Maneuvering in fits and spurts. First in were reports of what equipment was shut down. The reports were specific but the conclusion was sweeping; everything was shut down. Then followed reports of those parameters reading as expected. Professional discipline was bringing order out of chaos.

As soon as propulsion was lost, Control ordered Maneuvering to rig for emergency propulsion. This was the Throttleman's job. The Throttleman was also the Maneuvering phone talker during casualties. He wore a sound-powered phone mouthpiece on his chest and a headset over one ear. He was doing a heroic job of relaying the deluge of information coming into his headset ear while sending the outgoing communications ordered into his uncovered ear.

He was still the Throttleman and was performing his duties to rig the ship for emergency propulsion with one hand, the other hand keying his phone mouthpiece. The rig for emergency propulsion involved remotely disengaging the propeller shafts from the main engines and reduction gears, and making ready DC power from the battery to roll the shaft. Due to the bombardment of information, in which he was the middleman, the rig for emergency propulsion was slow.

Control had requested the status of the rig for emergency propulsion twice. The communication had gotten through, but the response had not been what Control wanted to hear. The Officer of the Deck became frustrated at the excessive time the evolution was taking. Wanting the Engineering Officer of the Watch's prompt attention, the OOD grabbed the 7MC microphone that announced directly into Maneuvering. Its speaker located directly above the Engineering Officer of the Watch's head, the 7MC would do the trick.

"Maneuvering, Conn. Report status of rig for emergency propulsion." The Engineering Officer of the Watch bristled at the noise above his head and at the interruption. At the moment he was a very busy man. He answered through his 7MC microphone.

"Conn, Maneuvering! We are in the process of rigging for emergency propulsion! I will inform you when it is complete!"

The Officer of the Deck in Control, Chief Warrant Officer Hammill, was an old hand and knew the young Engineering Officer of the Watch was in up to his eyeballs. The Officer of the Deck wanted to be supportive, but he had his own troubles. He responded calm as a blue sky.

"Maneuvering, Conn, aye. Maneuvering, Conn—the ship is sinking."

All eyes in Maneuvering zeroed in on the depth gauge mounted above the Electric Plant Control Panel. The needle was slowly drifting downward. Dead in the water, we were at the mercy of buoyancy. We were less like a submarine and more like a four thousand ton brick.

It was *Seawolf's* practice to run a little heavy. Slightly negative buoyancy made the submarine easier to control. When we moved forward through the water, our external control surfaces acted like wings on an airplane to control our three-dimensional position in the sea. We needed forward motion. We needed propulsion. We needed it now.

Caught up in our "engineering casualty," we had been reminded that we weren't the only ones with a problem. The Throttleman flew into motion and finished the last steps of the procedure.

"Sir. The ship is rigged for emergency propulsion."

Way was put on the boat and our depth was safely controlled.

The E- DIV LPO reported that the port ship's service turbine generator had suffered a fault in its voltage regulator circuitry. He ascertained this by opening the panel covering the regulator circuit to find a charred, still smoking electrical component. The malfunctioning component had caused the wild voltage fluctuations just before the generator's output breaker was tripped manually by the Electrical Operator to save the vital buss.

Number Two Motor Generator had tripped off line a moment later by the voltage transient affecting its regulator circuitry. Its regulator was "operating in the charcoal region giving a smoke ring output," as electronic techs liked to say.

Number One Motor Generator had been tripped off line inadvertently by the man sent to open the breaker for the now defunct Number Two Motor Generator. This was not known during the casualty, but determined afterward. This event was unfortunate, since having that machine running sooner rather than later would have truncated the direness of the situation.

Number Three Motor Generator was the only AC producing machine still running. Incapable of carrying the entire electrical load by itself, it was trying to do just that. It was keeping the AC busses energized with enough power to make the alarms and sirens try to sound, but not enough to give them full voice. It was also keeping the buss voltage just high enough to keep the emergency lights from energizing automatically. In its valiant attempt to do its job, it was inadvertently keeping us in the dark.

By now the darkness problem was being alleviated by E-division. Lights were robbed from unneeded places and given over to endeavors more valuable. Soon every electrical load possible was stripped off the vital buss by opening breakers.

Half the lights came back on which considerably lifted spirits.

Watchstanders not directly in pursuit of the problem were anticipating what would have to be done to make us well.

One sticking point remained. The electrical operator had observed vital buss voltage fluctuating wildly. The vital buss supplied voltage to the control rod drive mechanisms. It was possible that the control rod latching mechanisms had de-energized, energized, then de-energized again, causing the latching mechanisms to open then shut then open again. Instead of the control rods going to the bottom of the core unhindered, their travel may have been interrupted. This phenomenon—a ratcheting scram—could have damaged the control rod latching system. The control rods were the cork holding the nuclear genie in his bottle. Conducting a scram recovery with damaged control rods would be unsound engineering practice.

By now the captain was in Maneuvering. The Commanding Officer asked questions and offered suggestions, but left the Engineer to do his work. The Old Man knew that too many cooks spoiled the broth and this was a situation requiring good soup in the first go-round. We could not afford to muck this up.

Number Three Motor Generator was still plugging along. Number One Motor Generator was restarted. We got main coolant pumps back and the other items necessary to recover the reactor plant. What we now needed most was a critical reactor. After that, steam.

In Reactor Compartment Upper Level the RC division LPO and several other senior supervisors discussed the ramifications of a ratcheting scram. By the book—the Reactor Plant Manual—a ratcheting scram was a serious issue. Specific steps were required to verify the integrity of the rod control system before the reactor could be restarted. The Captain stepped into the conversation.

The RC LPO, Kenny Kohl, was a first class petty officer. The other engineering division LPOs were chiefs and senior chiefs. They began a discussion with the Captain about what had to be done before we could bring the reactor critical again. They quoted the Reactor Plant Manual, pointed out caveats and unknowns, talked about the paperwork that would need to be done to stay on the straight and narrow in accordance with the bible of Naval Reactors. After two minutes of conversation, the Captain turned to Kohl. Kohl was the most junior man present but he was the Reactor Controls Leading Petty Officer and therefore the resident expert by definition.

"RC LPO, what do we need to do to get my reactor up and running?" the Captain asked.

"Sir, there is a possibility that we had a ratcheting scram which means we should do some rod testing before we can start up," Kohl replied.

The captain shook his head. "We don't have time for rod testing," he said.

"We could do a modified short form pre-critical check off that would test the rod position indication," suggested the RC LPO.

"How long will that take?"

"Three or four hours."

The CO addressed the group. "Gentlemen, we are in a place where we cannot be found. We cannot raise the snorkel mast and use the diesels. I need that reactor critical ASAP."

"Give me better options," the CO said. Kenny Kohl had a ready response.

"Sir, we could verify the switch lineup and perform a Fast Scram Recovery. We'll keep an eye on the rod positions to make sure they operate as expected. If there is any hiccup, we scram the reactor and regroup."

"How long will the switch lineup take?"

"It takes fifteen minutes and I've had a man on it for the last five minutes."

The captain did not hesitate. "That's what we're going to do," he said.

Ten minutes later we began a fast scram recovery. A few minutes later the reactor was critical. Calculations showed the ship's battery, which had been supplying all our power for the last hour or more, to be near depletion. Time was running out. We needed steam to get the electrical generators rolling.

The steam plant had cooled and steam pressure was too low

to spin the turbines of the generators. The Ship's Service Turbine Generator that had been down for maintenance was ready for action again, but steam pressure had to build up before the turbine generator could be spun.

It was a race against time. Would the battery die and de-energize the reactor machinery before steam came up to a useful pressure? If that happened, it would go dark again. We would be worse off than before. Our options would be more limited.

We didn't choose to think about that. We chose to make sure it went right the first time.

Attention now focused on steam header pressure. Its gauge needles were rising at the speed of a glacier. The starboard ship's service turbine generator was now spinning, but without enough horsepower to be electrically loaded. Loading the generators too soon would stall the turbines.

John Parrott finally coaxed enough speed out of a turbine to attempt electrical loading. It worked. The generator began producing electricity even though steam pressure was still below normal value.

A year later, John Parrott would have a lively discussion with a Westinghouse engineer who declared it impossible for that steam turbine to carry electrical load with less than 180 psi steam pressure. Parrott would disagree.

Turbines started spinning and steam pressure continued building to normal operating pressure. One by one the machines that dropped off line a couple of hours ago were brought back up. In a couple of hours more, things were nearly back to normal.

Twelve hours later the faulty component in the port turbine generator had been replaced and that turbine generator was back on line. So was Number Two Motor Generator.

Emergency flashlights had their batteries replaced and were secured into their red holders. The ship's battery was fully charged. Submarine life continued as before.

Everybody debriefed the incident to figure out exactly what had occurred when. An impressively thick report was written for delivery to NavSea08, the Naval Reactors gods, who required their tithe of paperwork.

As soon as the danger was past the jokes began.

"Did you see Parrott? He was laying on the thing! Were you tweaking that turbine Johnny boy, or were you making love to it?"

Parrott laughed. "I was saying, 'come on, baby, you can do it… it's me, your favorite Mexican…do it for me…'"

Guys tried to ping on Clem Davis, the Throttleman during the incident. "A little slow on the rig for Emergency Propulsion, weren't you?" Clem was unfazed.

"I don't know what you're talking about. I had it under control." The guys guffawed. Clem laid the argument to rest.

"Did we sink? No. Did we have propulsion? Yes. I don't know what else to tell you..."

When asked how he knew it was the turbine generator regulator that had gone bad, Chief Lethin said, "I followed my nose. I opened up the panel and looked at the stinky, charcoal-looking thing, and my outstanding troubleshooting skills took over. I said, Ah-HA!"

They even tried to zing me, the Reactor Operator. "How high over the limit did you go on that startup rate?" Hey boys, I was right on the limit, not one smidgen over. I did offer that at such a high startup rate the reactor handled like a Ferrari.

"Oh yeah?" chimed in Jeff Parkin, the Electrical Operator during the incident. He had been standing next the Reactor Plant Control Panel when I was conducting the Fast Scram Recovery. "It looked to me like you were getting wild with that startup rate." I reminded him about parallax. He was looking at the meter from way over on the side. I was looking at it straight on.

"Sure, whatever you say, Wild Man." A nickname was born.

For many years afterwards, I remembered the darkness I experienced that day. I knew intellectually that in and of itself darkness is nothing. Darkness is a lack of, exactly opposite an abundance of, light. I had been in the woods at night, I knew the dark. What I didn't know until then was the complete absence of light. Intellectual contradictions aside, I could feel it. The darkness was heavy and thick. It smothered. I felt the warm sensation of vertigo in reverse. Rather than being dizzied by a height, it seemed there was nothing holding me up. I was free floating in a black void. When the Throttleman switched on a flashlight the sensation had vanished.

Back at Mare Island a month later an awards ceremony was held for the crew. This was a standard Navy occurrence about twice a year. Individuals so deserving were awarded various commendations, Navy Achievement medals, Good Conduct medals, Sea Service ribbons, and other military accolades. These awards were sometimes automatic, as in a Good Conduct medal which signified four consecutive years of not getting into trouble. Higher awards, a Navy Achievement Medal for instance, were often political. A career guy needed points to make rank and merit awards might be the only thing to put a 4.0 guy above a field of 4.0 sailors.

You could tell when the lifers were helping other lifers advance their careers with a marginally deserved award when the men the awardee worked with looked around during the reading of the citation, then clapped unenthusiastically. Perhaps another person was more deserving of the award, but this man had stood in line and done what he was told. His turn at the trough had come at last.

After a mission, scuttlebutt made its rounds regarding who was being considered for what medals. Every time the subject of awards came up I heard "he should get that, he earned it," and I also heard "if they give that fucker a Navy Achievement Medal for doing his job, I'm flushing mine."

Most of the time the awardee had actually made some significant contribution, going above and beyond the call of duty to repair a critical piece of machinery or computer system or whatever his specialty. The guys he worked with said, "Yes, way to go, good job, well deserved." Applause was hearty.

It did seem—in my unofficial reckoning by casual observation—that the big medals always went to proclaimed career men. My saying so will make some readers mad, but it was hard to believe that the guys with the cleanest uniforms did all the work.

Number Three Ship's Service Motor Generator was awarded a plaque engraved with a citation for valor. It had been the only electrical generator that stayed on line during the whole fiasco.

Honor well-earned and well-deserved.

A year later I was in the Crew's Mess relaxing with a cup of Joe. A couple of guys in quals were discussing their remaining signatures. Only designated system experts or people with supervisory qualifications were allowed to sign off level of knowledge requirements. John Parrott's name came up as an easy signature. I started paying attention to the discussion.

"What makes you say that?" I asked. The men had the impression that Parrott was less knowledgeable because he was so easygoing. They thought he didn't know much because he didn't hammer them for not knowing. The implication was that Parrott didn't know much so he couldn't ask much. I knew Parrott was more interested in helping people learn than he was in breaking balls.

Plenty of qualified people drew satisfaction from belittling non-quals and liked making those junior to them suffer as they had been made to suffer. This is one of the lowest character traits of the military.

These men had come aboard *Seawolf* after the Loss of All AC Power. They didn't know the whole story. They couldn't understand the chain of events without an hour-long monologue. I gave them only this:

"I don't know whether he's tough on checkouts or not, but he knows the plant. When the pucker factor is high off the charts, and you're thinking you might not see daylight ever again, you want John Parrott in the Engine Room."

Once again, honor well-earned and well-deserved.

34

Paradise Calling

Smoking to excess, drinking intoxicating liquor, and associating with bad women can ruin your bodily health.

— *The Bluejackets' Manual, 1940 edition*

AFTER RECOVERING FROM "Loss of All AC," the rest of the mission was uneventful. When the Captain made the announcement that we were done, cheers rang out. The submarine made a U-turn and for the first time in weeks, a Full bell was rung up on the Engine Order Telegraph. We cheered again.

Unapproved facial hair needed to come off before we made landfall. Guys took their time in the endeavor, trimming a little each day. One day the beard was shaggy, the next day neatly trimmed. The following day it became a goatee with muttonchops, then a Fu Manchu or a soul patch. Finally all faces were clean-shaven with the exception of a few mustaches. Trimmed to military specifications, of course.

Out-of-spec haircuts were corrected. Some had shaved their heads or gotten an uber-short buzz cut before going to sea. Now grown out, the uneven scruff on their head begged for trimming. Those who had not gone the "cut it all off and let it grow back" routine had two options.

The Chief Torpedoman set up an ad hoc barber shop in the Torpedo Room. During certain hours he would accept all comers at "Copeland's Buck and Cut." The Chief was an electric clipper man but his blades needed sharpening. His customers yelped each time he withdrew the clippers from their scalp. Those hairs not cut were pulled out.

The other option was reciprocal haircuts with a friend. He cut yours, you cut his. All my life I'd watched in mirrors as barbers cut my hair; how hard could it be? In the Stern Room laundry I wielded the comb; I thrust and parried with the scissors and tried not to butcher Keith Lyly's hair. Mostly I failed, but congratulated myself for not botching the job too badly. Now it was his turn to cut my hair.

Traveling southward from our operating area, temperature in the engineering spaces steadily increased. The boat's electricity and propulsion were produced by steam turbines. The turbines required thousands of pounds per hour of dry saturated steam. Eight-inch-diameter steam headers roaring with 500 degrees Fahrenheit steam may sound inviting in frigid waters, but we were headed to Hawaii and that's a long way from the Arctic.

The steam piping and components were insulated but insulation doesn't make things cold, only safe to touch, theoretically. Everything near steam got hot, pipe insulation included. The heat was attenuated as it passed through the insulation, but still radiated into the atmosphere of the engine room. Engine room temperature rose from ski lodge warm to sweat lodge hot.

Steam wasn't the only source of heat in the boat. Nearer the equator the ocean warmed. Cooling systems on the submarine ultimately used the ocean as a heat sink. Heat was removed from spaces and components where it was unwanted and moved to a place where that heat was unobjectionable—the surrounding sea. Heat exchangers for various systems were cooled by the Auxiliary and Main Seawater systems. As the sea warmed the effectiveness of the heat exchange dropped. Everything in the boat became warmer.

The Engine Room Upper Level watch stood under a 6-inch round outlet vent, wearing it like a hat. His hair blew straight down, as if he had just climbed out of a swimming pool. He held a white ceramic coffee cup full of ice.

I was going forward to the crews mess and asked him if he wanted an ice refill. He didn't hear me over the roar of air from his vent. He wouldn't leave the cold blast to inquire.

"What?" he shouted at me. I waved him off.

He nodded solemnly and held the cup of ice to his forehead. Offering to refill Engine Room Upper Level's ice was a token courtesy. After all, he had his own ice machine.

Resurrected and repaired from a shipyard junk bin, the ice machine had been installed outboard of the Number One Ships Service Turbine Generator one night. The ice machine was unauthorized and we kept it quiet. The engineering crew knew about it as did most of the engineering junior officers. Surely the engineer knew, and possibly the XO. Executive Officers seem to know everything. The ice machine wasn't brought up in conversation. If it did come up we played dumb. We were good at that. We were professionals in the secrecy business. Never allowing ourselves to be in a position of denial, we simply did not confirm.

> FORWARD GUY: "Is it true that you Nukes have an ice machine in the Engine Room?"
> NUKE: "Have you seen this ice machine?"
> FORWARD GUY: "No."
> NUKE: "Who told you such a thing?"
> FORWARD GUY "Ralph Lamaski."
> NUKE:"Lame-Ass is a Torpedoman, what the hell does he know about the Engine Room? You should be more careful where you get your information. Bad intel makes you look silly."

Every statement was true. No lies were told. Plausible deniability ruled. This is the covert ops biz at its best.

Before departing for this mission we had gone to sea for a week to test our equipment. Our squadron commander, Commodore Jack Maurer, came along. Well liked by *Seawolf's* crew, the Commodore had been captain on another SubDevGruOne boat, *Parche*, before he moved into senior management of the Dev Group. All nuclear submarine commanding officers were Nukes long before they became COs. We fondly referred to the Commodore as "Cap'n Jack."

A few years earlier, Cap'n Jack had snaked *Parche* through the Bering Strait submerged on their way to do some spying on the Russians. The passage through the Strait is split by two islands, Little Diomede and Big Diomede. The National Oceanic and Atmospheric Administration chart of the era shows passages on either side of the islands to be equally narrow with some depths at no more than 25 fathoms or 150 feet for you non-sailor types. It is reported that from the Tin City U. S. Air Force facility on Alaska's Cape Prince of Wales peninsula, that Russia is visible across the Bering Strait with the naked eye during the clear skies part of the year. Passing through the Strait with a submarine as big as a football field is reminiscent of Marko Ramius flying his submarine through the undersea canyons in the Hunt for Red October. However, I believe Cap'n Jack performed his tense maneuvers at a crawl. National security at risk was no time for showing off. I heard that on its return from the mission, great scrapes were visible down *Parche's* hull from as-close-as-it-gets encounters with Arctic ice.

Leaving the Bering Strait behind, *Parche's* crew took her into the Chukchi Sea, over the top of the world via the Arctic Ocean, and back "down" into the Barents Sea to do some proper spook work near the Soviet fleet headquarters near Murmansk. They did a good job too.

Of course, I knew none of those details at the time. I gleaned that information from books published long after my End of Active Obligated Service.

The Commodore preferred to hang out in the Engine Room. I think he liked the bustle and the noise of the Engineering spaces as opposed to the stoic sixty-eight degree calm of the Control Room and its adjacent spaces.

When the Commodore was aft, it was common for Nukes to lean on the diesel workbench and listen to sea stories. His stories were great. It was a pastime enjoyed by both sides. Also, my many bosses wouldn't fuss at me for not being more productive when the Commodore was holding court.

The Engine Room was warm, the Commodore looked thirsty. Someone disappeared and returned a moment later offering a fresh cup of ice water to the Commodore. He graciously accepted it.

"Thank you," he said and contemplated the cup. Fresh ice water was an unlikely convenience in the engine room of a nuclear submarine. The Commodore was a smart guy and a leader of men. He never turned around to look at where the frosty cup had come from.

"You got a...?" he said, gesturing over his shoulder.

"Yes Sir, we do."

The Commodore nodded his head. "Make sure it's gone before the ORSE team comes onboard."

"Aye, aye Sir. We will. You can bet your star on that."

The Commodore grinned. "I probably am," he said and began another sea story.

The Commodore had not come on this mission. We had been at sea eighty days and were now pointed toward Hawaii. Conversations centered on the first thing we planned to enjoy on dry land. Women were high on that list, but not as high as one might imagine. Food was a popular topic, especially fresh fruit and crisp greens. Any activity out of the sauna of the engineering spaces was worth an honorable mention. Men eagerly anticipated cool dark movie theaters, snorkeling in the cool water of Hanauma Bay, chilled salad bars and shrimp cocktails, and of course, icy-cold beer.

Skip Lethin, the Electrical Division Leading Petty Officer, the E-LPO, planned to ask a bartender for two bottles of the coldest beer available. He wanted to drink the first while watching condensation trickle down the second.

Jim Stanton, the Leading ELT, wanted a hundred pounds of fresh pineapple. What he really wanted was all that pineapple served by the wedge, one wedge to a Mai Tai.

Kenny Kohl, the RC-LPO, was planning a foray to several strip clubs he had visited before. We spent hours discussing the charms and limitations of each establishment. You recall that I said women weren't number one on the list, which was true, but they were still on the list.

A nautical mile short of the Pearl Harbor entrance a small boat met us and transferred aboard an admiral in sparkling summer whites. Introduced by name over the announcing system we looked blankly at each other. His name meant nothing to us. Then he was introduced as Commander Submarine Force, United States Pacific Fleet.

Well, why the hell didn't you say so? We didn't know Rear Admiral Whatshisname. To us the man's name was COMSUBPAC. Him we knew. He was the big boss of all submarines in the Pacific Ocean. Over the ships announcing system the admiral congratulated us on the mission.

"You have done a great service for the security of your nation" he said. "You reflect great credit upon yourselves and on the United States Navy." The platitudes were routine; we could quote him before he spoke. These guys plastered on a goofy smile when talking to the little people and they always said the same stuff.

Uncle Sam doesn't like free thinkers in charge. It seems to me that by the time a guy makes admiral, all imagination has been beaten out of him and replaced with the bureaucrat's need for keeping the politics right. That was my opinion at the time, and more so now; coupling my amassed knowledge of humans to Navy Times articles reporting the latest firings of commanding officers.

The admiral's tone was jovial. He spoke enthusiastically, with a touch of television game show host.

"As you hit the beach here on Oahu for some well deserved R&R, bear in mind that seventy percent of the hookers in Honolulu and Pearl City have herpes and thirty percent have tested positive for HIV. Welcome to Paradise."

Viewed on a map Pearl Harbor vaguely resembles a scrawny shamrock with the harbor entrance as its stem. Navy piers were to the right but always marching to a different drum, *Seawolf* turned left.

We tied up to a wooden pier in Pearl Harbor's West Loch; a banyan tree overhung our stern. This section of the Pearl Harbor military complex had been an ammunition storage area during WWII but now resembled a ghost town. Weathered warehouses stood on coral sand; unkempt grass had not seen a lawnmower in some time. Parked cars insinuated human presence, but no activity could be seen at this out of the way branch of the harbor.

I climbed out of the stern hatch into the Hawaiian afternoon. Breathing tropical air in the sunshine, I felt reborn.

A tarp was rigged over the Projects hatch forward of the sail. The Projects Weenies unloaded their treasures packed in nondescript cardboard boxes. A whistle periodically signaled the presence of a Soviet satellite in the visible horizon. All work stopped. The armed Marines guarding the operation faded into the trees or under a roof. People on the pier hid under the tarp until "all clear" was blown. The Russian snoop gone, work resumed.

A radioman, Sparky Foster, arrived on the scene driving a white USN pickup truck. I didn't know where the truck came from. Sparky shuttled sailors to a small bar a half-mile away.

The bar was a crummy little place, filled with scarlet and gold Marine memorabilia, but it had the two things we wanted most at the moment; cold beer and a pay telephone. Truckload after truckload of *Seawolf* men arrived at the bar in the back of Sparky's field-hand taxicab.

This was summer 1986. The cellular phone would not be invented for another ten years and we weren't waiting for it. The telephone booth enjoyed a brisk business. The bartender ran out of quarters. Those sailors carrying credit cards were paying for their buddies' phone calls with the card and taking beer in payment. One guy held a place in line for another to fetch fresh beers.

When my turn came the telephone handset was hot and sweaty. Many hands had cupped the talking end close to shield private conversations from the other guys waiting to whisper sweet obscenities to their own wives or girlfriends.

The bar was small and we its only customers. The bartender gave last call at 2200 hours, setting off a minor uprising. Forty sets of unfriendly eyes zeroed in on the barman. He protested that the place always closed at that time, it was the damn rules, not him.

The Captain stepped up and said, "I'd like to keep this place open a while longer." He opened his wallet and started piling twenty-dollar bills on the bar. Fortunately for us the rules evidently left room for discretion. When the pile of twenties was sufficient, the bartender relented.

It's a rare occurrence when a man can save his own life and make money at the same time.

We ordered more pitchers of whatever crap beer the barkeep was selling us. It was wet and cold and burned our throats. We savored every gulp. After being at sea with no booze for three months we were cheap dates. One pitcher of beer gave three men a warm glow.

We were nearly complete. Our mission accomplished, we stood on land in the warm tropical evening under a Hawaiian sky shimmering with bright stars. We were breathing sweet air and drinking cold beer. The only thing lacking was the company of the fairer sex. After being isolated from the world for the past three months, any female—eighteen to eighty; blind, crippled, or crazy—would have been a welcome addition to the party. Yet, even lacking women, life was still good.

The next morning we made ready to sail back down the channel into Pearl Harbor proper. Every crew member's head felt caved-in except one. Our resident non-alcoholic Electricians Mate Second Class Bill Greiner normally gave no one grief over their drinking; he did his thing and let others do theirs. But this morning he reveled in his sobriety and found amusement in the suffering of his fellows. Bill was a nice guy who would never abuse anybody and was, therefore, a popular target for practical jokes. It was only fair that he had his moment of triumph. Besides—we were already conspiring against him with a future prank.

The reactor had remained critical throughout the night. We were still steaming and making electricity but the propulsion system had been shut down.

The Engineering Watch Supervisor gave the order.

"Engine Room Supervisor: Bring steam into the Engine Room. Warm up, equalize, and open Main Steam Three and Main Steam Four. Warm up the Main Engines."

The ERS repeated back the order verbatim and issued appropriate orders to the Engine Room watches.

Main Steam-3 and Main Steam-4 were hydraulically operated valves on either side of the Main Steam system. If a steam leak occurred downstream of either MS-3 or MS-4, the valve could be slammed shut to isolate the leak. High pressure, high temperature steam running wild in the people space is very bad for business.

The big valves were located overhead. Smaller, hand-operated bypass valves bled steam around the shut isolation valves to heat the downstream piping and components. The bypass valves were also overhead.

Every component in a steam system is subject to high heat and moisture. Sound engineering practice dictated that petroleum lubricants should not be used on valve stems in high pressure steam systems. Oil would just burn away, leaving a mess. The best practice was to keep the valve stem clean. Steam valve stems, the part that had to be rotated to operate the valve, were often "sticky." The movable metal parts rusted to the non-movable metal parts making the valve difficult to operate. That the bypass valve had been shut at 500 degrees Fahrenheit, and has since cooled to room temperature, added to the problem. The materials used to manufacture the different parts of the valve expanded and shrunk differently with temperature changes.

These issues combined to render hand-operated steam valves difficult to operate. The stubborn ones tended to stick on their shut seat. Muscle was required to pop them loose. Cheater tools, ad hoc pry bars and the like were strictly verboten in Admiral Rickover's nuclear propulsion program. Manpower is more than a word. It's a definition.

When Tony Rezetti, the disheveled, not-enough-sleep, hung-over Engine Room Upper Level watchstander, reached over his head to open the first bypass valve, he found it stuck shut. Tony was a stout fellow and had plenty of muscle to do the job, but it was a chore to call upon those muscles this morning. He was also on his tiptoes to reach the valve, making it that much harder to get a firm twist on the valve hand wheel. He gritted his teeth and leaned into it. The valve popped loose and Tony opened it fully.

The bypass valve on the other side was stuck too. Hands over his head, straining at the valve hand wheel, his face turned red with effort, his eyes flew open wide. He let go of the hand wheel and slowly reached around to gingerly probe the seat of his pants.

"Oh no," he muttered in disbelief. Last night's revelry had come back to haunt him beyond a simple headache. Three pitchers of beer and no supper had turned the content of his guts into soup. Applying all his strength to the stuck valve had drained energy away from other bodily functions normally expected to operate on automatic. His sphincter relaxed only for a moment but once that isolation valve was cracked open, pressure in his churning guts did the rest.

His eyes downcast in abject misery, he spoke to the Engineering Watch Supervisor.

"Request permission to be relieved for an emergency head call."

The EWS glared at him. "Are you kidding me? We are right in the middle of engine room startup. You can't take a break!"

"But Chief," said Tony, wide-eyed with distress. "I've got a problem…"

"You have many problems Petty Officer Rezetti, but I don't have time to list them," said the EWS. And so, Machinist's Mate Second Class Tony Rezetti started the Engine Room and manned his watchstation while *Seawolf* sailed from West Loch to Sub Base Pearl. He made the trip with a lot on his mind and a load in his pants. Welcome to Paradise.

35

Hound Dogs

Occasionally a person of immoral habits succeeds in joining the naval service. It is your duty to report immediately any suspicion or rumor of lewd, lascivious, and scandalous conduct.

— *The Bluejackets' Manual, 1940 edition*

LOOKING BACK, IT seems odd that I never heard sailors on liberty say 'I heard this city has a great art museum,' or 'where is the theater district?' Men on the submarine hailed from every corner of the United States and from all social stratum. One would think some culture might have somehow leaked into the crew but apparently not. The main pursuits ashore were beer to drink and women to ogle.

Married guys were as enthusiastic about women as single men, albeit with different outcomes in mind. Although not a universal practice, a few wedding bands stayed behind on the boat during forays into town. Some rings came off the minute we left Mare

Island—to let the white scar fade—but not all. Most rings stayed on fingers and married men stayed married, in practice as well as by definition.

I'm certain wives back home worried about their man's fidelity, he being out of sight and out of reach. Husbands definitely worried about what was going on in their absence. The friendly single guy who lived next door, the attentive check-out clerk at the grocery store, the waiter who lingered; all gave speed to the hurricane whirling in a man's head months away from home.

"On the beach" wearing their "glad rags" married men traveled together for mutual support and rescue from feminine temptations which would surely lead to anguish and alimony. Rather than violating their vows, married guys' intentions were more voyeuristic than horn-dog. After enduring months with a crowd of men using the F-word in every sentence, a woman's laugh was pleasure itself.

A breath of perfume from a woman passing by could stop a conversation. Just for a moment each man was transported to another time and place known only to him. Lounging in a nightclub near a group of women talking and laughing was an evening well spent. It made a man feel normal again.

Pressure affects people in different ways, but pressure inside a man is always cumulative. Men are much the same as volcanoes. If pressure is not periodically vented, disaster is sure to follow. Especially single men, whose yearnings are mostly hoped for or imagined wonders, the way children anticipate Disneyland.

The passions of married men become more measured through experience; banked coals in place of a raging fire. The object of their amorous intentions is clear and they learn patience, either by force of will or through love. The first time a married man goes to sea he will be faithful because it's the right thing to do. The second time he goes away, he is no longer startled by opportunity. When anticipation of his home fires aren't keeping him warm is when trouble may start.

"Steve, my man, that woman at the far end of the bar keeps looking at you."

"Knock it off, I'm married."

"I'm sure that's what your wife is right now telling your next-door neighbor down at the No-Tell Motel."

"She's not like that."

"I hear you, big guy, I hear you. I'm sure she's a good girl. But she's still a woman alone, and that changes things. At the moment, she is geographically single, just like you. You know the old saying, 'There are no married men in a foreign port.'"

"I have steak at home, why would I want a hamburger in Hawaii?"

"Because your steak is 3,000 miles away and a tasty hamburger is sitting at the end of the bar..."

Steve remained faithful to his wife. So did most of the married men on the boat, nevertheless, these twisted rationalizations of geography and opportunity vexed many married men.

Morals notwithstanding, a sailor's desire for women was encapsulated perfectly by National Geographic. In a documentary about plains animals of the Serengeti, the narrator explained that "the duty of every male in the herd is to mate with as many females as possible. The duty of every female is to mate with only the best males." Although the program specifically addressed African antelope, it also explained men and women in two tidy sentences.

A million years of DNA is hard to deny, yet women did not trump our quest for alcohol. Months at sea without alcohol was proof we weren't addicted to the bottle; at least, we weren't when we first made landfall. Sailors with a drinking problem on shore discovered that an alcohol-free ship on an extended voyage is an effective detox center, albeit not a nurturing one.

More than simply a source of alcohol and entertainment, bars and nightclubs represented a 180-degree change of structure. Establishments where alcohol was served provided unabashed revelry and *joie de vivre*, a healthy dose of feeling like an average Joe replacing the tight professional containment of the boat, even if just for an evening.

Beer was good, better was where beer and women coincided. In the opinion of some, the best combination was cold beer and naked women.

Strip clubs were fueling stations for young hormones. If you couldn't yet grow a decent mustache, you could still bask in the heady mixture of booze and come-hither women in a fantasy world of sexual suggestion. These women understood sales better than any Harvard MBA. They were expert at their value proposition which was making young men feel special and vulnerable to the feminine wiles on display. The atmosphere of the place was enough to keep a man digging into his pockets.

Separating men from their cash is a strip club's business, and business is good when the Navy is in town.

Every group of men leaving a strip club has that one guy who is convinced that one of the strippers likes him. After all, she smiled at him in a special way as he tucked his sweaty dollar bill into her G-string. To the gentleman's chagrin—or relief—liaisons between patrons and dancers are as rare as lottery winners. These ladies make their living mining the miners, not falling in love with cash-poor sailors.

Actually speaking to a woman in a more genteel social environment, a nightclub for instance, was emotionally risky. The unfired clay of a young man's self image translated a simple "No, thank you," as confirmation that he was fundamentally unacceptable. On the other hand, the thin smile of the stripper welcomed him without judgment. The clear boundaries of stage and depth of pocket saved him from the embarrassment of being poor. His dollar was a good as the next man's. When his pocket ran dry, he moved to the back of the room and continued to enjoy the show financed by the front row.

Given the environment and the moment, throwing wadded dollar bills onto a stage worn by high-heels and bare knees seemed appropriate manly behavior. Hooting while a bored woman disrobed to a heavy metal soundtrack with his besotted buddies cheering him on was safer to his ego than risking rejection from a "nice woman" in the bar at the Shore Bird.

By noon of the second day in port the entire boat knew the best happy hours and which joints had the hottest strippers. The timing of our arrival coincided with a dental assistant's association convention at the Waikiki Hilton. Sailors reeking of Old Spice were invariably found at the Hilton bar, even though their wallets preferred happy hour at Moose McGillicuddy's or The Crow's Nest. Packs of sailors watched for groups of unescorted ladies. Lone hunters looked for potential; a glance, a shy smile. Even an imagined hint of encouragement signaled a likely prospect to cut out of the herd.

Women on vacation without their significant other can be as predatory as men. Determined to have a spicy story to tell their girlfriends back home, Mainlanders on vacation were better odds than a local girl. Married women left on their wedding bands without shame. Even more endearing, they never complained "Why haven't you called?" They did not want their vacation memories to cause trouble back home. This is what has been reported to me by those who know.

36

Waikiki Nights

Be careful, however, that you do not pick the occasional "shirk," "piker," or "fourflusher" for your friend, as such a man will invariably get you into trouble...

— *The Bluejackets' Manual, 1940 edition*

WE MOORED *SEAWOLF* to a pier in Submarine Base Pearl Harbor, brought on shore power and shut down the reactor. A nuclear reactor is not shut off in the sense that you shut off your automobile and walk away. *Seawolf's* pressurized-water reactor was shut down by driving the control rods in to the bottom of the core. The control rods absorbed thermal neutrons which interrupted fission. Without that chain reaction, the reactor was not much more than an inert assembly of exotic metals. Even so, a Navy reactor is monitored continuously for the life of the ship.

Relieved of the watch and changed into civvies, I crossed the brow and stood on the island. I stood on concrete, but it was land.

It didn't move underneath my feet. It was Hawaii. It was spiritually renewing to climb out of the boat and into the sunshine even if only temporarily.

I marveled at the water sparkling in the late afternoon sunlight. Pictures don't do anything justice, I thought.

When the word "aquamarine" was conjured, the person doing the naming must have been looking at tropical water. This Hawaiian water was iridescent green-blue and clear. Several feet underwater I could see small fish cautiously approaching the submarine's hull and then darting away as if they smelled something they didn't like.

Although the industrial "Haze Gray" of Navy was everywhere, palm trees broke the outlines of ships and buildings. Inland, lush green hills rose against the bright blue sky. Vivid colors were rendered surreal by the warm breeze.

I thought about surfing the big rollers on the north shore or climbing Diamond Head. Equipment could be rented through the Navy Exchange. I mentioned these ideas to Jim Stanton and Rob Kersch.

"Are you crazy? It's Happy Hour." So we climbed into a taxi and said "Waikiki."

A popular tourist destination, the district of Waikiki was jammed with shops, restaurants, and hotels. Bars lined a white-sand beach crowded with sunbathers. Mainlanders were easy to spot. Their bare skin glowed white. I winced at the sight, anticipating the vicious sunburns that would ruin their romantic evenings. The sun's rays were intense here and the cool breeze deluded one into believing that they were tanning comfortably. Tomorrow they would be bright red and slathered with Aloe Vera.

Our first stop on this night out was the Hale Koa Hotel. The Hale Koa was reserved for Department of Defense personnel. That was us. The hotel's Barefoot Bar was just the place to start an evening out.

After a few drinks there, Rob, Jim, and I roamed Waikiki. Everything was walking distance and the night was warm. We visited the Down Under Bar. It was one of many bars with a city front on Kalakaua Ave and a back door opening to the beach.

We found ourselves at the Crow's Nest above the Jolly Roger restaurant. A show was going to start in an hour and we had luckily found great seats at a table near the front window overlooking the street. The waitress asked if we'd like another drink before Happy Hour ended. Jim asked how many drinks we could order at one time. "I can only stack them three deep," she said.

We agreed that would be fine. The waitress brought nine Mai Tais made with fresh pineapple. Jim was well on his way to eating a hundred pounds of pineapple wedges.

The show was hilarious. Two musician/comedians played guitar and banjo, and they sang original songs. They called themselves The Blue Kangaroo. Each set became raunchier and raunchier. People in the crowd would toss out insults which inspired new bawdy songs. I was never sure if the songs were prepared in advance or made up on the spot.

The evening over, we jumped into a taxi and went back to the boat. The next day I ran into Jim in the crew's mess. "How you doing?" I asked. My head was buzzing and Jim did not look chipper.

"I think I have pineapple juice poisoning," he said glumly.

"Too much pineapple juice can bring a man down," I said. "It probably didn't have enough rum in it." Jim agreed that must have been the case.

My duty section had the duty that night, which was a good thing. I needed drying out. The next afternoon, as soon as our divisional work was done and liberty was put down, we went straight back to Waikiki.

At the corner of Kaiulani Avenue and Prince Edward Street stood a bar; the Rose & Crown Public House. Now gone, it was a lively watering station and a favorite of English and Australian sailors. Plenty of Yanks were in attendance, both military and civilian, but Australians gave the place its flavor. Their accent was fun to hear. They laughed full-on without pretense or holding back and they were friendly. What's not to like about a guy who called me "mate?"

Submarine sailors respect a good insult and foreign colloquialisms were favorites. At the Rose & Crown our vocabularies expanded. Since the Brits and the Aussies both spoke a version of English, we could understand most of their conversation but not all.

"Where did you just come from, mate?" an Aussie asked.

I said, "Moose McGullicuddy's, over on Lewers Street."

He knew the place. "Oh, yeah," he said. "There's some lovely cock over there."

That startled me. "Whoa, hold on there buckaroo," I said. "Maybe you and I don't play for the same team."

"No, no, mate! Girls, women, you know... pussy," the Aussie grinned.

Why the term "cock?" I asked. He pointed out that that's what a man used, with any luck, in the situation to which he was referring.

"That's a turn of phrase you Aussies ought to rethink," I said.

He bought me a Foster's and we ogled women together. Male bonding knows no borders.

It was the non-American, non-politically correct phrases that made us laugh the hardest. American sailors reserved particularly rude or ambiguous pejoratives heard from other places for use on the unknowing. More than one junior officer has been startled to be asked for a fag by a leering petty officer enjoying the other man's uncertainty. In the Royal Navy the request for a fag would result in a cigarette being proffered. In the U.S. the term connoted an altogether different concept.

The Australians and the English had a love-hate relationship which, when mixed with alcohol, occasionally erupted into shouts and trouble. One regular topic for insult and injury between Aussies and Brits was the Queen of England. Many a brawl had its origins in a rude comment about the Queen. British sailors didn't seem particularly concerned over the honor of their Sovereign; in fact, it seemed to me they regarded royalty as useless theater. On another occasion they would likely refer to that same Queen with the same unflattering terms.

What English sailors did welcome was any justifiable excuse to have a go at those "convict gin jockeys" from Australia. Aussies talking trash about the Queen fit the bill well enough for a Saturday night.

During these disturbances the police were never summoned. Uprisings were efficiently quelled by a squad of Samoan bouncers. As a group they resembled an NFL offensive line, nothing was getting past them. Their vigilance gave the impression that you were constantly being sized-up. And, of course, you were being sized-up. By the time you were three steps through the door, the bouncers already knew how they would take you down if you started something.

These large men were sufficiently impressive that the drunkest chucklehead heard the warning clanging inside his head: "Do not screw around with these guys."

The Rose & Crown was not a rough place *per se*; it was simply populated with boisterous young men of different nationalities, many recently returned from sea. In their youthful primes they were loaded to the gills with the world's most dangerous cocktail: one part alcohol, two parts testosterone, shake and serve without garnish.

Before this mission I had come to Oahu for Electronic Technician's Maintenance School. That had been my first time at the Rose and Crown and that trip had been entertaining at an epic level.

Two of my ETMS classmates and I sat at a table in the middle of the room with a pitcher of beer in front of us and a covey of pretty women at the next table.

A back door whipped open and a group of Australian sailors crashed in. This clearly was not the first stop of their evening. Laughing and shouting they roared down the steps into the bar. The last one in, a little guy about five-foot-nothing short, stumbled on the last step and nearly fell. Well into his cups, he whirled and glared up at the colossal bouncer standing his post against the wall.

"'E tripped me!" cried the little sailor. His friends laughed.

"No 'e didn't mate; you're just too drunk for stairs," his buddy chided. "Come on now, let's have us a drink." The feisty sailor did not take his eyes off the bouncer. The Samoan looked down at the diminutive Australian like a sleepy lion watching a gazelle.

"I can take 'im," Feisty declared and began to peel off his jacket. Behind him his buddies groaned: Oh, for fucks sake.

"Somebody else do it. I did it last time," one of the Aussies said. Another man stepped forward, grabbed Feisty by the shoulder, and spun him around. The fist started low and traveled upward in a swift arc to intersect Feisty under the chin. It struck with a loud pop, lifting the smaller man to his tiptoes. When Feisty's weight touched back down, his legs were unwilling to support him. He folded onto the floor in a heap.

The puncher took hold of his unconscious friend by the shirt collar and dragged him across the floor. The puncher smiled at the bouncer.

"Sorry for the trouble, mate.'E gets this way sometimes. No offense meant." The puncher propped Feisty into a sitting position against the wall next to a table where the other Australians were already seated. The group proceeded to down pitchers of beer leaving an empty glass ready on the table.

Across the room we worked on our own pitchers, alternating our attention between the Australians and the group of young women chattering in Italian at a nearby table. We hoped for a floor show from the Australians. We were also hoping for conversation with the Italian women. Each one looked like a fashion model and they seemed far more sophisticated than we. We were intimidated.

We held a conversation young men have shared since the first cave man was brained with a mammoth bone for rudely plying his affections.

"Just say, 'Hi,'"one of my classmates suggested.

"I'll go invite them to come over and sit with us," I said.

"Listen Einstein, they haven't spoken a word of English," Johnson said. "How are you going to ask them anything?"

Thompson and I wore wedding bands, Johnson was single. Still, none of us was looking to score or do anything with these women except bask in the glow of their company.

"The brunette just looked over here. She's checking us out."

"They're all brunette, dopey. They could all be Sophia Loren's stunt double. They're looking over here because you're staring."

"Watch the Australians. They are going to start something, sure as hell."

"You order another pitcher and I'll go sit with those women."

"They seem to be having a nice time. We should just let them be."

Our insecurity now rationalized, we ordered another pitcher. We were grown men, twenty-five years old and not prudes, yet we were quibbling like schoolboys over women too beautiful to talk to. By the time I had worked up enough nerve to approach the Italian women, they were paying their bill. As they left they flashed beautiful smiles at us and said, "Ciao." We were crushed.

Fortune favors the bold and we had flown the colors of the timid. I vowed to never again hesitate.

We had hoped for more entertainment from the Aussies, but they had kept to themselves and didn't cause the ruckus we were expecting. Our fantasy women gone, I approached the Aussies.

Indicating the man slumped against the wall I said, "He hasn't moved. Did you kill him?"

The Puncher grinned broadly, "Naw, 'e' just needed a little rest."

I stepped closer to look at Feisty propped against the wall, his eyes shut. A gentle snore greeted me.

"Had to deck 'im to save 'is life," the Puncher said. He indicated the Samoan bouncer and spoke with respect. "'At's a big fella."

When we left the pub a couple of hours later, the Australian sailors were gathered around the upright piano. They were singing Waltzing Matilda for the enlightenment of a patron who was regretting his confession of never having heard the tune. Feisty was with them, in full voice, waving his pint like a choirmaster while the bored Samoan bouncer looked on.

37

First Impressions

Save on Command Presence! Qualified orders over $25 ship free

— Banner Ad on web browser following INTERNET search:"Quotations on Command Presence"

"WANT A DATE this evening, boys?" she cooed, stepping into the circle of light from the street lamp. She and her female companion were slender and dressed provocatively.

Two buddies and I were walking toward the lights of a bar up the street in Pearl City. Chris was quick to seize the opportunity and threw his arm around the woman's shoulders, bringing her in step with us. He knew where her conversation was headed and made a preemptive strike.

"Ladies, spend the evening with us and we won't even charge you!" he said cheerfully. Both women laughed.

"No, silly man, you gotta pay us. We show you a good time," one giggled. Her accent suggested Asian, but I couldn't have said if it was Thai, Chinese, or Philippine, or anything else for that matter. I can tell you that her voice was low and sexy. Her laugh jumped out, easy and light.

Chris indicated two men walking twenty yards behind us. Both had the fresh-faced look of teenagers. Physically they were diametrically opposed. One was wiry, with a bounce in his step far too cocky for his modest stature. The other was round, and puffed to keep up.

"What about my friends there. How much for you to show them a good time?"

She looked over her shoulder. "No," she said. Chris persisted.

"C'mon, how much for the big guy?" Her answer left no room for further discussion.

"Not for any amount of money."

We had been politely trying to lose the younger guys following us. They wanted to hang out with us but they were not our crowd. They were Forward Pukes, non-Engineering types, and worse, Nubs.

A submarine organizational chart is divided into departments and divided again into divisions by job requirements. People naturally congregated into these work groups in their off-hours, like families living in the same neighborhood.

You hang out with those you know.

It's not uncommon to see Nukes and Forward Pukes partying together, but the two guys following us were some of the youngest on the boat. Since they were going into bars they were probably twenty-one but seemed fifteen. Their conversations had an adolescent quality that caused us to roll our eyes and plot ditching them.

A couple of hours earlier they had found us in a bar close to the Pearl Harbor Sub Base where the boat was moored. They found the place the same way we did, following the scent of beer and the glow of a neon Budweiser sign.

The bar was filled with Marines, many in uniform. Even in civilian clothes Marines stand out. The traditional high and tight haircut along with their ramrod-straight bearing made them easy to spot. We sailors stood out like three sheep among a pack of wolves.

We occupied a booth against the wall when the two younger men spied us. Grateful for familiar faces, they crowded in. Jim shot me a disgusted look that said, "We should have taken a smaller booth."

The Forward Pukes were Schumacher and Twinkie. Schumacher was "Shu" for brevity, and Twinkie had been nicknamed for his resemblance to a snack cake. In his personal battle of the bulge Twinkie's position had been overrun. On his first underway he had packed on enough mass to tighten his poopie suit until the brass zipper up the front looked as if it might burst at any moment. People avoided looking at him directly for fear of losing an eye when the inevitable event occurred. He continued to add to his real estate until he was placed in the Fat Boy Program.

The Navy had taken a stand against overweight personnel and Twinkie's height to weight ratio was out of spec.

I'm sure the program had an inoffensive official name that was positive and inspirational, but I don't remember. We called it as we saw it. It was a mandatory diet and exercise regimen designed to trim pounds from overweight sailors. It was a program for fat boys. The Fat Boy Program.

Submarines travel in a hazardous environment and are at risk every moment they are at sea, regardless of Defense Condition. A submarine sailor is in many ways similar to big city firemen with a major exception. If the fire gets too hot, there is no outside to run to. When the feces strike the air circulation device, the last thing anyone needed was some fat bastard complicating the situation with a heart attack. Besides, who could carry a big man through watertight doors and up ladders?

Everyone on the boat was aware of Twinkie's Fat Boy status. This was a support mechanism of a twisted sort. Not emotional support, not even by the most imaginative definition. The man was chided and derided routinely. Everyone not on the Fat Boy list became a self-proclaimed weight-loss expert. If The Twink looked at a piece of cake too long someone was sure to smack the back of his head.

"Don't even think about it fat boy," they would say. "And get those mashed potatoes off your plate." Nice guys would say these things. Hey, we weren't trained in psychology.

Gaining weight at sea was easy. A meal was served every six hours. Most meals were an "all you can eat" affair. Chow was served family-style with platters of food at each table. Serving yourself an extra scoop of mashed potatoes was easy, as was taking an extra steak. Pies and cakes were made fresh every night. Ice cream machines churned all day. A man was perpetually surrounded by the temptation to eat. The people tank was only 300 or so feet long and 20 feet wide. Burning calories was near impossible. If you did a jumping jack anywhere on the boat you would likely split your head. There was a treadmill in the Projects spaces that could be used only when the Projects Weenies weren't doing their Projects thing. One lone piece of exercise equipment, available occasionally, was not enough to fulfill the physical fitness needs of a hundred men.

On my first underway I eased my belt out an inch or—well, let's call it an inch. On subsequent underways, I limited my intake so that Levis left at home would still fit when I returned.

A man trying to get a weight problem under control on a submarine at sea needed Herculean willpower.

Meanwhile, back in Pearl City...

Seated in the bar booth with the three of us, the young guys began to feel comfortable. Too comfortable, given the environment. With half a pitcher of Coors Light in them, they began making disparaging comments about Marines, joking about "jarheads" and "grunts." These jokes were common Navy rhetoric, but location is everything and this was most definitely not the place for pinging on the Marine Corps. If their apparent sense of invincibility was founded on we Nukes being present, these two lads were going to be disappointed.

As they drank they got louder. Their comments attracted attention not in our best interest. We told them to knock it off, but it was too late.

I thought the lights went out when a figure loomed beside me and placed large hands flat on our table. As he bent down to face level his biceps looked about to split the Marine chevrons on his shirt sleeve. A similarly proportioned mountain stood at his side.

"You guys got a problem with Marines?" the sergeant asked with a tone that caused my mouth to go dry.

Jim indicated the Nukes at the table and said, "We have no problem with Marines." Pointing at the Forward guys he added, "These two hate Marines. You can have them for dinner if you want."

Twinkie and Shumacher sat motionless, looking as though they had just been lobotomized with a coat hanger.

Camaraderie has it limits. Jim was not afraid of a fight and would have taken on the whole place for a ten-dollar bet, but by his reckoning Forward guys were sacrificial. From time to time you'll lose a few, but no worries about that, there was lots of them.

But the Marines had no concept of Forward and Aft squids. The Marines considered them and us as a set. Our immediate future was looking grim. That was when Senior Chief Kelly stepped through the front door of the bar.

Senior Chief Kelly was one of a small group of Projects guys on the boat whose job requirements included being amazingly physically fit. He had skills I was not cleared to know about. He also ran *Seawolf's* physical fitness program, including the Fat Boy Program.

Observing the entire barroom's attention focused on one table, he quickly grasped that it was not looking good for team Navy. Stepping between the Marines, the Senior Chief reached up and placed a firm hand on a shoulder of each.

"Gentlemen, these guys are friends of mine. Is there a problem?"

The two Marines studied the Senior Chief for a long minute. At five feet nine inches tall, the top of his head was level with their shoulders. However there was nothing in his demeanor that suggested he was the least bit intimidated. In fact, his was the upper hand.

The Senior Chief was built like an Olympic athlete. Close-cropped sandy hair, a leading-man square jaw, and muscular shoulders tapering to a trim waist made him look like a recruiting poster. If you melted him down he wouldn't render enough tallow to make a birthday cake candle. At the last Command physical fitness test he had zipped off a hundred push-ups and a hundred sit-ups in less time than it takes a grown man to shave.

His khakis were well tailored and pressed with sharp military creases. The Marines' eyes lingered on the service ribbons above the Senior Chief's breast pocket. The array of medals represented there were the real deal. A Silver Star, a Bronze Star, Legion of Merit, the Navy and Marine Corp medal, and others. These were combat medals, medals of valor, warrior medals. They were not the standard cluster of "thanks for showing up" medals worn by most of us in the peacetime Navy.

If these Marines wanted to eat this Navy Senior Chief for dinner, they should have brought more Marines. That he outranked both of the Marines was secondary in this situation. He was fucking impressive.

Fortunately for us, the Marines thought so too. His well-timed intervention had shifted the focus of the situation away from us. The Marines stepped back.

"No problem here, Senior Chief," said the first one. They were neither cowed nor insulted. These two Marines had never met the Senior Chief before, but they knew him. He wore a Navy uniform, but he was one of them.

Wolves don't eat wolves. Wolves eat sheep.

The Senior Chief leaned over our table and spoke quietly. "You guys get the fuck out of here."

It was good advice. This was no time for a childish argument about who said what to whom and who started it. A light had shown at the end of the tunnel and we did not hesitate. We skedaddled.

Walking up the road we encountered the two professional women plying their trade. After that failed attempt word got around that poor Twinkie couldn't get laid even for money. As if he didn't have enough image problems. As cruel as we were, I am amazed that he didn't snap and either punch someone, or curl up in the fetal position sucking his thumb.

I later heard that the Senior Chief had spent the rest of the evening drinking with the Marines.

You hang out with those you know.

38

Tough Guys

It is not flesh and blood; it is the heart which makes us sons and fathers.

— *Friedrich Schiller, Die Räuber, Act I*

UNLIKE SURFACE SHIPS, nuclear submarines do not replenish underway. A U.S. Navy surface ship on extended deployment is able not only to restock its pantries and fuel tanks from supply ships; it also receives and sends mail for the crew. In contrast, a submarine at sea has only those provisions that were onboard when it left port. Receiving and sending mail—traditional mail, snail mail—was out of the question in 1986 when I was at sea on a nuclear submarine.

A submarine's stealth advantage would be ruined by rendezvousing with another vessel or aircraft to trade mail bags. Other nations—especially the disfavored nation du jour—should not know the whereabouts of our submarines either on-route or on-station. Messages such as, "be at latitude-x and longitude-y at noon of the 25th" is just the sort of information an enemy could use against us. After all, that's how American naval aviators killed Japanese Admiral Yamamoto in World War II.

On a submarine at sea, letters written to your sweetheart were saved until you handed them to her yourself. I kept a diary. More than a recollection of the day's activities, it was a continuous letter to my wife. I shared whatever tidbits of the day's activities I hoped she would find interesting and told her about events involving men she knew. I wrote to her the latest news on when we might be home. This last seemed ridiculous since I would be home before she could read it. But that was the nature of my diary. Nothing held back; no entry too silly. It was the first truly vulnerable thing I had ever done.

More importantly this diary was a sappy love letter, detailed in long dream-state run-on sentences. Conjured from memories of good times we had shared and the hope of a bright future together, this perpetual letter was for me a sanity maintainer. Hopefully she saw it as endearing. I never asked.

When I returned from sea, she gave me a packet of unsent letters bound with yellow ribbon. We spent an afternoon locked away together, each reading the other's thoughts during the months we were separated.

Communications technology at the time allowed families to periodically send a telegram—a Family-Gram—to their man at sea. Limited to fifty words, elaborate codes were set up between sender and receiver. For example, if the house you were selling had sold the Family-Gram code might be "HS," which counted as one word. The message might say "HNS" if you were still the owner. Clever wives could send the Gettysburg Address using only fifty words of code. Some secret communications between husband and wife would likely have made an eavesdropping code breaker warm around the ears.

Incoming messages were received encrypted. The submarine rose to periscope depth and a radio mast poked up through the surface. Message traffic was downloaded in moments, much like a personal computer downloads a compressed file. The message was unpacked and decrypted. Printed out, it was delivered by hand.

Usually a Family-Gram was a personal message carrying significance only to the intended recipient. However, good news made its way quickly around the boat.

A new baby boy was announced to the entire crew over the ship's announcing system, congratulating the father and sending best wishes to mother and child. Included in the announcement was a cheerful admonition to raise the boy to be a good sailor. The announcement was followed by cheers and backslapping. Good-natured bad jokes were heaped on the new dad.

"Didn't know you had it in you, kid."

"Does he look anything like me?"

"How's your wife and my kids?"

The new dad responded sardonically, as a submariner must. "You guys are funny. Why don't you go take a nap in a torpedo tube?"

Comments made in jest often carried a cruel edge that unintentionally cut deep for men far from home. Young men, insecure by way of immaturity and inexperience, themselves only casually acquainted with fidelity, tended to hold one nagging doubt about wives and girlfriends.

"Hey, weren't we at sea nine months ago? Didn't Jones stay ashore with a broken hand? I hope your kid doesn't look like that fuck-knuckle."

The new father knew that this absolutely is not the case. However, on arriving home he will inspect the baby with an eye sharpened for detail. Afterward he will be quietly ashamed for allowing that faithless thought to germinate. Such are the tribulations of seafaring men. Even good news stirs phantoms in lonely minds far from home.

Family-Grams were screened; bad news was not delivered right away. Whatever the issue, you couldn't do a damn thing about it. The only thing receiving bad news at sea could do was make a man crazy and there was no situation in which that would be helpful.

Bad news was withheld from its recipient, to be delivered by the Chief of the Boat only when the boat steamed into San Francisco Bay on our way home. If a family tragedy had occurred while you were deployed at sea, you would not learn of it until you were almost home and the COB handed you the "Bad-Gram."

The Chief of the Boat didn't like to come aft. It was hot and steamy back in the Engine Room where the Nukes worked and the COB liked his uniforms crisp. When he did come aft after we passed under the Golden Gate Bridge, we watched warily from our watchstations.

"Uh-oh, here he comes."

"I was getting 'Grams every couple of weeks and then they stopped. That's not good. The COB might have one of those for me."

"He better not have anything for me."

"All I know is if he gives me one, I'm kicking somebody's ass." The speaker looked at the man on watch next to him. "Sorry bub, you're closest. It'll have to be you." The target of this pronouncement returned a bored look and made a stroking gesture with his fist.

Seeing the COB on this errand, my breathing went shallow. My youngest daughter, only six months old had gone through a scary bout of whooping cough just before this trip. I didn't yet have full confidence in her well-being, the poor little thing, but had pushed that anxiety out of my mind. Catching the COB's eye had unexpectedly brought those fears back.

Once he passed me by, a full breath brought me new life. I got away clean, but others did not.

"Uh-oh, Smitty got one. You know his dad's been sick."

"No, I didn't know that."

"Yeah, my wife and his wife are pals. I guess his dad hasn't been doing well for a while."

"You think…?"

"Yeah, he's reading it now. Look at his face."

Smitty's Family-Gram was definitely bad news. Men quietly moved away from Smitty, giving him what privacy they could, a few square feet in a crowded world. We went about our work, both real and imaginary and let him be. It was the most we could do.

Later, men walked up to Smitty and wordlessly put a hand on his shoulder. He received a few pats on the back, and other seemingly insignificant kindnesses. An act as simple as someone refilling his coffee cup without making the requisite smart remark was not random or trivial today. These small obtuse gestures were as much empathy as our maleness allowed.

We did not yet know for certain the source of Smitty's melancholy; we would know when he was ready for us to know. For now we knew enough. The man had gotten a Bad-Gram and he was bummed.

Coming back from the Mission of '86, we stopped at Pearl Harbor for a few days before starting the two-week transit to Mare Island.

Easter Sunday had passed. Families back home had gathered in a park on Mare Island and enjoyed the holiday. A picnic, complete with an Easter egg hunt, had been videotaped. We received the tape in Hawaii and made ready to watch it once we were underway.

The Crew's Mess was packed. Men were sitting or standing on every available horizontal surface. A carnival atmosphere filled the space as men jeered and jostled each other in high spirits. Finally the lights dimmed and the video began.

We laughed and cheered as our kids, dressed in their Sunday best found Easter eggs and romped in the California sunshine.

"Man, those deviled eggs look good."

"What I wouldn't give for a hunk of that barbecued chicken right now."

"Look at the salad." We constantly thought of foods we were doing without. No detail went unappreciated.

"Red tomatoes, oh baby..."

"You know those cucumbers are crisp."

Caught up in a fresh vegetable high, the crowd exhaled oohs and aahs. The video moved on. Men joked with each other.

"George, your oldest daughter is really growing up. Jesus, look at the— "

"Steady men, steady," the COB broke in. The words went unsaid but the thought was completed in each mind. A murmur of agreement and appreciation rustled through the mess. George's daughter was indeed growing up wonderfully. He knew it and it scared him.

"Are you letting her date yet George? I would be glad to be first, only as a favor to you, you know, to help her steer clear of the bums..."

"Yeah, thanks," George said. "You wouldn't live through my 'prospective boyfriend' interview."

On screen an exuberant ten-year-old boy jumped as high as he could for the camera.

"Hey, Tom, your son looks a lot like your neighbor the Marine!"

Sick of hearing that joke and beyond reach of the man, Tom pointed at the jokester and ordered no one in particular, "Hit him for me." The jokester grinned and winced under a half-dozen good-natured punches.

We groaned when one of the boys missed a catch in the softball game. We howled and high-five'd the father when a girl with pink ribbons in her hair got mad and threw a hard-boiled Easter egg at the boy tormenting her. Then the last section of the tape played.

Each family, minus husband and father, stood in front of the camera and gave a personal message to their man on *Seawolf*. Each wife or girlfriend gave some cheery news then told us they missed us and loved us. The message was directed to one, yet caused every man to feel closer to home. Love sent to one of us was love received by all of us.

An eight-year-old boy waved at his dad and wistfully asked, "When are you coming home?"

My little daughters Hannah and Sarah, angelic in Easter frills but unsure of the camera, waved and said, "I love you, Papa."

Except for the sounds of the video, the Crew's Mess was silent. Men who didn't smoke were trying to light bummed cigarettes. When the overhead lights came on men left quietly, each withdrawing into a private place inside. Many eyes were ringed red. The toughest faces, unashamed, made no attempt to hide wet cheeks.

Throughout the boat, a hush settled like a fog. The rough banter of men at sea was suspended; each man deep into his thoughts. Those who had been on watch and missed the video were curious about the abrupt change in atmosphere but allowed others their silence without judgment. Each sailor performed his duties with silent dignity, safely ensconced behind castle walls. Personal questions and comments went unvoiced. With time, lighter hearts and the natural humor of camaraderie returned. From then on we were more discerning about which videos we watched.

Nothing I could do would hurry the day when I would again see my wife and little daughters. Looking at photos of them taped to the wall of my bunk, I felt a warm glow. Their faces made me happy. But there was a limit to how much I could dwell on them. Thinking too deeply about what I was missing skewed my sense of comfort into despair teetering on the brink of overwhelming. I could not afford to go there.

In moments like this it was easy to regret enlisting just as I imagine a man in prison wished he hadn't committed the crime. It had seemed like a good idea at the time, now it merely seemed stupid. It was plain now that this Navy life was no life for a family. At least not my idea of family. The Navy would be a career of absentee fatherhood.

It wasn't as if I was working my way up to a position where my family would benefit from my years of service and advance to a more regular life. I was not a salesman, flying back and forth across the country waiting for the day I would be promoted to sales manager and stay close to home while the other guys traveled. There was no pot of gold for a Navy Nuke. If I stayed in the Navy there would be stints of shore duty, but my career as a submariner would always return me to sea.

Just as we all had placed our lives on hold for the duration of the mission, I also had to place my feelings and personal longings aside. It was as if there was a bureau drawer back at home in which I placed everything of me except the Navy's part. The stuff in that drawer stayed home with my family. On my return, I would open the drawer and put on all of me again.

In the meantime I had to get through it. Being a young man and yet only vaguely aware of my emotions, I simply turned down the volume until there was no sound in my head.

39

Certifiably Nuts

Certifiable: An honorable credential useful in many aspects of submarine life.

— The Submarine Sailors' Imaginary Book of All Knowledge

A SUBMARINE TRAVELS better deep. The same sea pressure that would kill us if it could also helped us travel faster. Deep water is cold water and colder is denser. As the pressure and density of the water increased, propeller slippage decreased. The screw became as close to one-hundred percent efficient as it could get.

By keeping average main coolant temperature on the low end of the Green Band, an extra turn or two of the propellers could be gained for any given reactor power. We called these "going home turns." Every turn of those screws brought us sixteen feet closer to home.

On the way home we drilled and we cleaned. We also practiced to speak as civilized men. We had become accustomed to rough language, so much so that it was difficult to change our poor speech habits. Keith Lyly helped me.

I began a conversation: "I can't fu—"

"Hey, hey, hey," Keith stopped me.

Right. "I can't wait 'til we get fu—"

"No, no, no," Keith again chided me. I got it now. I completed the thought without vulgarity, enunciating each word like I was learning a new language.

"I can't wait 'til we get home."

Jim Boivin had been spending time in the log room and discovered some Navy supply catalogs on microfiche. He filled out an order form to procure a .50 caliber machine gun. I asked if he thought that was a good idea.

"It's a Navy supply catalog and I'm Navy property. It's just one piece of property ordering another piece of property. Besides, it'll be fun. We can go to the shooting range and blast away with it," he said.

Never mind that someone would surely ask what a submarine needed with a machine gun and quash the whole business. Or all the trouble he would get into if it actually was delivered. When he signed the delivery receipt, a pair of bracelets with a chain could not be far behind.

But that was logic and we had been at sea a while. Logic had become less and less important. I chose another tack.

"Have you thought about ammunition, Jim? That thing will go through fifty-cal rounds like Chivas Regal at a Kennedy family reunion."

"Good point," he said. "I'll work on that." He went back to studying microfiche. I hoped that would slow him down until he got bored with the machine gun idea, but I would learn that at sea, men latch onto an idea like a terrier on a pant leg. They'll leave it only when pried away by brute force, or having their attention shifted to a new interest. I have seen men spend an entire mission filing and sandpapering a three-quarter inch threaded nut into a ring to wear on his steel dust stained finger. I believe it was the Mission of '85 when someone whiled away the weeks by polishing the Wilton machinist vise bolted to the Diesel Workbench till it shone like chrome plating.

His name was Michael Bernhardt, Electronics Technician Second Class, one (each), and we called him Bernie. Even engineering junior officers slipped in their formality and called him Petty Officer Bernie. Bernie had some Japanese in his heritage and liked to wear his beard in a Fu Manchu that would have made any Samurai proud. At sea he wore a headband carefully folded and knotted behind his head. On its front was the Rising Sun. Bernie like to stir things up.

A forward guy came into Maneuvering on some business for the Engineering Officer of the Watch. While his paperwork was being looked at, he stared at Bernie's headband. The forward guy was a visitor to Nukeland and knew he was off his reservation. Still, he decided to cross that social boundary and speak his mind.

"What are you doing wearing the Jap flag?" he inquired with no humor in his voice.

"I like it," Bernie replied evenly.

"You know they bombed Pearl Harbor," said the Coner. Jesus, the man was serious, he thinks he's informing us. This should be good. There was nothing like a matching of wits for quality entertainment. Everyone in Maneuvering waited patiently. We didn't have television, but we did have live theater.

Bernie looked at the man quizzically. In a voice rife with innocence he said, "We were in Pearl Harbor a few weeks ago. I didn't see any bombs." Bernie turned to the Throttleman. "Did you see any bombs?" The Throttleman covered the beginnings of a grin.

"I didn't see any bombs," he said. The Throttleman looked at the Electrical Operator. "Did you see any bombs Rudy?"

"Cut it out," interjected the forward guy. "It was a long time ago goddammit! You know what I'm talking about. World War Two." The jig was up; he knew history. The man persisted.

"You ought to take it off. That headband is bullshit. This is an AMERICAN submarine, in the UNITED STATES Navy. That," he pointed to the headband, "is un-American."

"Un-American?" said Bernie. He was beginning to heat up. "I'm from Texas, how the fuck could I be un-American? It's a statistical impossibility. I'm as American as apple pie and baseball. I'm as American as Audie Murphy. He was from Texas, too, you know."

"Who's Audie Murphy?" our forward guy asked.

"Look it up," Bernie said. "Get back to me when you know something."

"I still don't think it's American," said the forward guy as he left Maneuvering. He was serious in his opinion. Bernie called after him, shouting his *coup de grace* down the engine room passageway.

"I'm as American as a black velvet painting of Jesus, Elvis, and John Wayne holding hands for eternity!"

Point, set, and match.

When first reporting to *Seawolf,* I noted messenger buoys built into the hull superstructure. Flush with the surface of the weather deck they were the size of a bathtub; one up forward, one aft. They could be released from inside the boat when a last resort was all we had left. They would float to the surface tethered by 1200 feet of stainless steel rope that would guide rescuers. A stainless steel plaque mounted on each read:

"Below this buoy lies USS Seawolf (SSN-575) with 125 men. Contact the U.S. Navy or the U.S. Coast Guard."

Sweet Jesus on a whole-wheat cracker.

Never mind the arithmetic problem of 1200 feet of cable when we were in 20,000 feet of water. This was a classic submarine practical joke laid on us by the designers. Yet, who could know, we might be lucky enough to sink on the continental shelf along the coast of California. It's possible that we could bottom in less than 1200 feet of water.

I wondered if the fisherman who found the buoy floating would be running late to pick up the kids. Hopefully on his way home he wouldn't just pass it by as floating junk. Hopefully he would take a minute to make the call. Cell phones wouldn't be invented for another 10 years; would he have change for a pay phone? Would he reach a bored government employee who would put him on hold until he said, "Hell with it" and hung up? Would he forget the whole thing until he heard about it a week later on the evening news?

Oh, was I supposed to call someone? Perhaps he would rifle his pockets for the scrap of bait wrapper on which he scribbled the telephone number. Is that a four or a nine?

Readying for my first sea voyage I observed shipyard workers welding steel bars over the messenger buoys. I anxiously informed the first Navy person I saw, my sea dad, Bob Cranker. He showed no alarm.

"Yeah, so what?"

"The messenger buoy won't work!" I protested. "If we are stuck on the bottom no one will ever find us."

Cranker laughed. "See there, how smart you are for a new guy!"

That was my welcome to Special Projects as well as to the diabolical humor of Bob Cranker. On our way back from Hawaii, I got another dose of Cranker's wicked wit.

Uniform regulations were relaxed at sea. As sailors are prone, we began pushing the edges of the envelope. On our way home, we were getting cocky. Small gestures of anarchy made us happy. Not enough anarchy to get a guy in trouble, just enough to feel we were our own men.

Personal dosimetry, a must for everyone on a nuclear submarine, was being worn in other than prescribed locations. Instead of Thermoluminescent Dosimeters being worn on the belt as decreed by good engineering practice, some TLDs were attached to shoelaces or strapped to the bicep. TLDs hung on chains around necks like a talisman.

To correct this deficiency, the command put the word out via the Plan of the Day. The POD stated that all personnel must wear a belt, the dosimeter to be placed on the belt at the body's midsection.

Men had also been pushing the limits on footwear. At sea we wore sneakers. A couple of guys had broken down the heels of their canvas tennis shoes and were wearing them as slippers. The POD announcement put a stop to this practice, stating proper foot gear should be worn that encompassed and protected the feet and did not readily fall off.

Bob Cranker studied the grammar of the order and decided it said something different. He strolled into Maneuvering to take the watch wearing only shoes, a belt with a TLD on it, and not one thing more.

Word got around and the Executive Officer popped into Maneuvering. He shook his head slowly when he saw Cranker sitting at the Reactor Plant Control Panel in the buff. Bob had a bath towel between him and the seat, more for comfort than for hygiene.

"You can't be serious," said the XO in amazement.

"Sir, I'm as serious as a reactor scram at test depth in a jammed-dive with a 30-degree down angle and torpedoes in the water."

Cranker showed the XO how the Plan of the Day had said "proper foot gear," and "a belt with TLD." It required nothing more in the way of clothing.

"Be careful out in the Engine Room," the XO admonished. "Those steam pipes are hot."

The next day the plan of the day carried a clarification to yesterday's announcement. It was far more specific about acceptable attire.

Since our previous Engineer had attempted suicide, the Navy decided we needed watching. A Navy psychologist had come on the mission with us to evaluate and determine who of us was on the verge of cracking up. I'm sure he also got a professional ticket punched by going to sea for ninety days. That's pure speculation on my part. Perhaps Navy psychologists get all the sea time their careers require. Perhaps.

Machinery that does not work is not welcome on a submarine. It takes up space and uses oxygen without contributing. It is despised by submariners. In keeping with the submarine tradition of not mincing words, we named the psychologist Doc No-Load.

He was a friendly guy, Lieutenant Commander Something-or-Other. The Shrink. He was often referred to by his Indian name, He-with-No-Job.

He loitered at various times in various places presumably observing various people going about their various duties. I never saw him take notes or obviously pay attention to anyone or anything in particular.

He reminded me of a guy I knew growing up who was a masterful hunter. The hunter's head remained motionless while his eyes scanned the landscape; side to side, top to bottom, near to far. I sensed that like that hunter, this psychologist didn't miss much.

Doc No-Load had crow's feet at the corners of his eyes. Those only come from squinting into the sun or laughing. Since he lacked the leathery face of a man constantly in the sun, we rightly judged that he had a sense of humor.

He had a tough gig, the Shrink. He was an outsider, observing a bunch of closed-up smart asses who knew they were being observed and would rather fuck with him than be supportive. Even so, after a few weeks at sea our suspicion of him faded. He would come into the Engineering spaces and joke and talk about everyday things just like we all did.

O-gang (the officers) traditionally held a wardroom Scrabble tournament. Teams were chosen among the officers and the competition was hot. The tourney normally took a week or so. After four weeks hard at it, a winning team was finally declared.

Amazed at the duration of the tournament, O-gang was even more amazed when Doc No-Load emptied his pockets on the playing board. He had shanghaied half the vowels.

"You did better than I thought you would," he laughed. "Especially with only four E's and four A's."

Open talk of tortures that would have made the Inquisition seem tame rumbled around the wardroom. But the moment passed and the joke was on them. Everyone laughed and went on with underway life.

Doc No-Load was not invited into the poker tournament.

Months later, when the mission ended we sailed into San Francisco Bay and John Parrott joked with Doc No-Load.

"Well sir, you've been watching us for months. What do you think? Are we crazy?"

The Navy psychologist became serious. "I can report that you are all one-hundred percent certifiably nuts," he said. "But there is nothing wrong with any of you."

High praise from He-with-No-Job.

40

Home Free

Home Free: The final surfacing of a sea deployment.
Almost Home Free: Learning that you have duty the first night in.

— *The Submarine Sailors' Imaginary Book of All Knowledge*

THE DAY BEFORE we were to arrive home, a member of the ship's Welfare and Recreation committee came through the engineering spaces selling chances on the bridge pool and the anchor pool. Five dollars bought you a space equal to 30 seconds on the clock face. The winners would be determined by the official 1MC announcement saying we were "passing under the Golden Gate Bridge", then "the Richmond Bridge," and finally "the ship is moored." Each block of time bought had a potential value of $150. More than $150 worth of chances was sold; any overage went into the Welfare and Rec kitty. The Man running the show always gets his cut of the action.

We all knew this was the last mission for *Seawolf*. When the boat surfaced to enter San Francisco Bay we were happy because we were coming home. We were also aware that surfacing signaled the completion of the final dive of the old boat. The final surfacing marked the submarine equivalent of a perfect record, a dive to surface ratio of 1:1. One surfacing for each dive. Anything less would have been total failure. Submarines either have a perfect record or they don't come back.

The surface of Golden Gate Strait had enough chop to give us a couple of roller-coaster ups and downs as we came in, but nothing like the first time I had gone out that way. The Officer of the Deck reported over the announcing system that we were passing under the Golden Gate Bridge. The crew cheered. Our passage smoothed abruptly. We were in San Francisco Bay.

The engine room hatch was opened. Men stood on the deck underneath and gazed up at blue sky. Fresh air poured in and we sucked it in. Men under the hatch were compelled to move aside to give others a crack at the sunshine. The ladder, which had been stowed for sea, was broken out and hung underneath the hatch. Now men crowded up the ladder, poking their heads up out of the boat. It was dangerous to have people topside when underway, the deck being narrow and wet and the vessel moving swiftly. Only those necessary to operate the ship were allowed up top. But that didn't stop us from peeking.

It was a bright summer day in San Francisco, the 23rd of June, 1986. The sunshine, typical for that time of year and routine to the locals, caused us to wince and shield our eyes. We sailed past the same cityscape I had seen photographed from Sausalito except that now it was much closer. We passed San Francisco only a couple of hundred yards off her waterfront. People on the promenade waved. I wonder if they had any idea of how enthusiastically we waved back.

After my turn on the ladder, I ran back to my bunk to dig sunglasses out of my belongings. Fool me once…

Seawolf docked in Alameda, across from our old acquaintance the *Carl Vinson*. Remembering our last encounter a few of us were anxious. We did not know if our escapades 105 days previous would return to haunt us.

While torpedoes were offloaded, forays were made to the gedunk machines at the head of the pier. Orders for sodas and candy bars were filled for those on watch who could not go themselves. The pay phones at the head of the pier sprouted lines of sailors waiting their turn. Mostly the calls said the same thing; we'll be home in a couple of hours.

Antsy to be on our way, we finally settled our accounts in Alameda. We untied and backed away from the pier out into the bay. Once again answering bells in the forward direction, we pointed the bow toward the northeast corner of the Bay.

The left turn required to enter the Napa River from the south was tricky for the submarine. The current running out from Benicia Bay past the southern end of Mare Island was swift. The Napa River was on our nose with a current of its own. As soon as the boat began its turn to port, the current caught its length and tried to swing it sideways. The propellers and the rudder being at the far end of the lever, the boat was unwieldy once this phenomenon had taken hold. The best method was to not let the current get hold of the ship. This was done by keeping more speed on than a cautious skipper might choose, then making that turn abruptly and charging up the river.

This maneuver required steady nerves in the man on the bridge. This last trip in, one of the junior officers had the bridge to get his qual signature for bringing the boat into port. He backed off the bell too soon and the boat began to swing.

The bow swung past the centerline of the river and a moment later was aimed at the sea wall just above the Coast Guard station. A small crowd was gathered there; family and friends trying to glimpse the boat sooner than the folks lined up pier-side at berth 19.

The JO on the bridge realized his error and ordered more speed. It took a moment for the order to be rung up on the Engine Order Telegraph, a moment for the Throttleman to open the Main Engine throttles, and another moment for the turbines to respond. Suddenly spinning faster, it took a moment for the propellers to gain purchase in the water and another moment to overcome inertia and put more way on the ship. These moments combined meant that we turned too far and accelerated too late. We were now driving straight for the crowd on the seawall.

A three-hundred eighty-seven foot long submarine does not turn on a dime. Neither does it reverse directions like a professional hockey player. The bridge tried to helm the boat over to starboard, but we were fast running out of river. The bank was coming on quickly.

The people on the seawall looked agitated. Smiles gone, they sidled away from the boat's line of travel, then they scattered. This was going from bad to worse.

Someone on the bridge, probably the Captain, stepped in and rang up All Stop. That was followed by Back One-Third on the starboard main engine and Ahead One-Third on the port main engine. Ten seconds later that changed to Back Emergency port and starboard. Behind the boat, water churned in a white froth. The boat's motion in the wrong direction arrested, the length of the hull was slowly and ponderously levered over to starboard on the hinge of the rudder. Once again we were in the center of the channel.

We steamed up the river to berth 19. A waiting tug tied on and nudged us sideways to the pier.

We were home.

Wives, girlfriends, kids, moms and dads, and a set of grandparents dressed in their Sunday best, lined the pier.

The unhappiest men were those who had duty tonight. They had to stay on the boat one more night before they could go home to sleep in their own beds with their wives. So close and yet so far.

Lucky to not have duty, I threw my sea bag into the car and climbed behind the wheel for the twenty mile drive to Fairfield. Robin laughed because I was driving 45 mph on the freeway. Everything seemed to be moving so fast.

For too long I had been looking no farther than thirty feet inside the boat. Now the horizon blurred off into the distance. It would take a few days for my depth perception to return.

We passed a car that I was unfamiliar with. What's that car? It's a Hyundai. A what? It was new, just out, a new brand of car.

We drove past a movie theater marquee listing a movie I had not heard of; Purple Rain. Prince? Prince of what? An American prince? Oh, a performer named Prince. Is that his first name or his last name? His only name, oh…

Small revelations such as these occurred for a month. It was amazing how much seemed to have changed in the three and a half months I'd been gone. More than gone, I had been out of touch with the world and everyone in it. Our own little world had been defined by a steel tank, three-hundred eighty-seven feet long and twenty-eight feet around.

My little girls kept looking at me as if they hadn't seen me before. Maybe I had changed. They surely had. When I had left, baby Sarah was crawling. Now she wobbled happily on unsteady legs until she sat down hard. She was leery of me for a few days. I probably startled her by regularly sweeping her up and squeezing her. I covered her cheeks with kisses and I'm sure she looked to her mother for rescue thinking, "who is this guy…?"

Hannah at two years old was accepting right out of the gate. She had enough memory of me to be more enthusiastic of my return. She wanted all my attention and I willingly gave it. All, except for the attention I saved for her mother.

After Robin caught me up on latest developments of the girls, the colds, the flu, whooping cough, the skinned knees, and other minor household dramas, she asked how the past few months had been for me. She never said, but I'm sure she was disappointed with my answers.

What could I say? It was—fine. What could I tell about the day to day of the past few months that would even make sense to her? OK, see, I stood my watch, and Keith, and Rob, and Mark stood their watches, and Clarence, and the other Mark, and Mike, and Johnny stood their watches, and Jim did ELT stuff, and I ate, and I slept. I worked and I studied. I walked from my rack to the crew's mess and back to the engineering spaces. We ran drills. Some funny things happened, some jokes were played. Boys will be boys.

She must have found it tremendously un-fascinating. What could I say to convey my experiences during the last three and a half months? I told her what I could of our major casualty, but it now seemed lame. The lights went out, okay, big deal; any housewife knows to check the bulb. Did you blow a fuse?

What I told her most about were the stopovers in Pearl Harbor. Being out in town with my buddies that she knew made for the best conversation anyway. She could relate to that. I told her so little about life on the boat and so much about Hawaii that it seemed to me she had the impression that I had been off on a three and a half month vacation. I resented that, but didn't know what to do about it.

Coming home was wonderful, but I had expected more. I'm not sure how much more. I thought an angel chorus would be singing and a heavenly light would be shining down on our little bungalow in Fairfield. The rest of my life would be a breeze since I had done this marvelous thing for king and country. Doors previously closed would open and I would be welcomed.

"You went to sea on a submarine, on 'a mission of great importance to the security of the United States,' thank you, thank you! Please do come in!"

Instead, I simply returned to real life. The yard needed mowing, the garage was a mess. The eight-foot-tall rosebushes in the back yard needed trimming before they fell over and punctured someone. The car needed its oil changed and there was a squeak in the brakes. The hinges on the bathroom window were filled with paint and rust. They needed a thorough going over before the pane was broken pushing open the window.

Curiously, the rest of the world had not come to a standstill in my absence. *Seawolf* and her crew had, for all practical considerations as well as literally, disappeared from the face of the earth for 105 days. Life on the planet had continued without us. In fact, only a couple of pages of names on a telephone list had even known we were gone.

Having just returned the boat went into a period of "stand down." For a week, men came to work on their duty day and went home for three days after being relieved the next morning. Work had to be done, but otherwise, we stayed away from the boat as much as possible. After a week of maximum family time, the normal in-port routine started up where it left off when we had gone to sea. My light weeks were once again seventy-two hour work weeks, the heavy weeks ninety-six hours.

My family settled in with me as the new guy in the house. Hannah and Sarah forgot that I had ever been gone. Robin did not forget. Things had changed with us.

Our garden untended, weeds had grown.

41

Plank Owner

War is hell. Peace is insufferable.

— *Military axiom*

ON JULY 8, 1986, Hyman George Rickover, the father of the nuclear navy, died. I'd never met the man; only saw his back disappearing into an office when I was at the S1C prototype on the occasion of a royal visit before he retired. Yet he was such an integral part of nuclear submarines that I was a little sad on hearing the news. The legendary admiral was gone and my submarine was next in line to be stricken from the record.

Scheduled for decommissioning, *Seawolf* was thirty years old and had lived through massive changes in submarine technology. Many of those changes had been surgical. Her time up, her services no longer required she was being retired, like it or not. Into dry dock she went, an old woman going into hospice never to leave alive.

The decommissioning process was planned to take nine months. The reactor would be defueled. Equipment that could be used on other ships or submarines would be stripped off. Every tank and piping system was to be drained, dried out, and sealed. Electrical systems would be de-energized and de-humidified. *Seawolf* was to be embalmed.

Theoretically, if war broke out tomorrow she could be reactivated, refueled, and her systems replaced in less time than it would take for a new submarine to be built. This was the basis of the mothball fleet.

In actuality, the military world had evolved past those old concepts. It would be impractical to bring back from the dead an old ship, to say nothing of the vast leaps in technology that would need to be retrofitted into an old platform. Mothballed was not inactivated. Mothballed was dead.

Mothballing was bureaucracy hedging its bet, too nervous to declare it over and done. That would wait a few decades when the same bureaucracy would finally say, "What the hell are we doing keeping this old hulk? Scrap it."

The decommissioning would be organized by the shipyard. Sailors—ship's force representatives—would do much of the work under the supervision of a Shift Test Engineer, an STE.

Major steps in the project had to be done in sequence. When something didn't go right, or a question was raised, it took time for the engineers and Naval Reactors Regional Office, NRRO, to ascertain the correct procedure. In the meantime, progress on decom came to all stop.

The Nukes spent a lot of time monitoring systems that were no longer working. We took reams of logs with rows of entries that read zero. It was our duty to monitor, and to log, and to store the logs. It was not our job to decide if it should be done.

At 2000 hours on a duty day, the Engineer Officer summoned Will Grant to his stateroom on the barge. The barge was positioned on blocks in the dry dock next to the submarine as temporary working space. The ENG's office was cramped. A fold-down writing surface bolted to the bulkhead and a chair filled the room.

Submariners are accustomed to tight quarters. We thought it was a decorating style. Women's magazines touted Elizabethan, or

Tuscan, or Early American. We had Cramped and Dismal in one color scheme, multiple shades of gray.

In the 1970s a Ford executive became infamous for complaining there was only $2 million budgeted to redecorate his office. He felt he needed $6 million to do the job properly. He wouldn't have done well in submarines.

The ENG's office had a high shelf festooned with miniature duck decoys. About three inches long, they were exquisitely carved and painted. A likeness of a Wood duck, a Canvasback, a Buffalo Head, and a Mallard were represented alongside other duck cousins. The ENG's desk was littered with tiny shavings from his Exact-o knife. Will was impressed. Perhaps this was why the Engineer was so easygoing. The man knew how to unwind.

The Engineer asked Will, as the Leading Petty Officer of Reactor Controls Division, how it was that RC Division was the only Engineering Department division that regularly turned in its quota of Divisional Training hours on time. The other divisions were constantly late, struggling to record enough hours. The rigorous schedule of the department performing major decom evolutions left little spare time to get everybody together in one place for training.

Divisional training was done when other work was not pressing. Therein lay the crux of the problem: other work was always pressing.

Will explained that he had communicated to all the men of RC Division what did and did not count as recordable training. Knowing the rules, anytime one man was involved in an evolution that could be counted as training, he assembled all the RC division personnel he could find. He converted a typical work procedure into an ad hoc training session. Every man in the division was doing the same. RC Div turned in training hours to spare.

The Engineer looked thoughtful. He said, "You know, that's not the Navy way." He was referring to the outdated Machiavellian method of "command and control" as the Navy way. Someone must give orders; others must obey. Working together as equals was not militarily correct. Someone must control; others must be controlled.

Will looked hurt. "Eng, we comply with all the training rules, everyone pitches in to get it done. If that's not the Navy way, it should be." On a roll, he continued his dissertation on the way it ought to be versus the way it was.

"The COB does a compartment cleanliness inspection every Friday morning to decide if he needs to call a Friday afternoon field day. Reactor Compartment upper level is always SAT. Want to know why?" The Eng was interested. Will kept going. "The space is always SAT because every day I tell my guys that when their work is done, they can go home, provided that I never hear from the COB about our space being dirty. The men take care of it themselves without being micromanaged."

Will had found his stride. "It's not necessary to hammer on people to get things done, sir. They are quite capable of managing these picayune household chores without direction. Or should I say, without oppression." Will would have been pushing his luck with most officers, but he trusted the Engineer as a man.

Letting people go home for the day when their work was done violated the principles of command and control. What if my boss needs something and if I don't have the manpower to do it, I'll look bad. Therefore, I will keep all my men here until my boss leaves for the day. Cover-your-ass was the rule of the day among most of the junior officers and chiefs.

Will had been called down by the RC Division Officer for sending his division home for the day during a ship-wide Field Day. The Field Day was called for a four-hour period. Will let his men leave when their space was clean, not adhering to the "clean for four hours" standard. To get a job done efficiently you set performance standards, not minimum time requirements. A task required to be completed in one week always takes a week to complete, even though it's a half-day task. Any business school graduate knew this principle.

Will made the guys work until the space passed his inspection. Then he set them free. Will knew what the COB looked for in his inspection because he paid attention. Reactor Compartment Upper Level always passed the COB's inspection.

Because they were smart enough to leave quietly, no one from the command realized that RC-DIV men were already on the beach by the time their space was officially inspected. Other divisions on the ship dawdled for hours until they were released. They polished the same valve again and again, for the required four hours.

When the RC Division Officer discovered his charges had gone home early on a Friday afternoon he was furious. As a Naval Academy graduate he was brought up to believe that enlisted men should be watched at all times. He considered pressing charges against Will, but the personal implication was too great.

Will Grant was relieved of his position as Reactor Controls Leading Petty Officer. He was clearly not the Navy model of the guy in charge. The command didn't want consensus in a work group. They didn't want people working as a team. They wanted firm discipline, the Navy way. Someone needed to be inconvenienced to make sure those in-charge stayed in-charge. Inconvenience must take place; hate would be fine too, as a bonus.

Will exited naval service soon afterward. RC division never ran so smoothly again.

There's nothing as miserable as a peacetime military. For a submarine, in-port was peacetime. At sea we did our jobs, overcame problems and emergencies, and grew our beards.

In port, the most important issues were haircuts and shoe shines.

Finally all the machinations of decommissioning the submarine were complete. The ceremony was set for March 30th, the thirty-year anniversary of *Seawolf's* commissioning. In her first year, the President of the United States had taken a short cruise on her. That President was Dwight D. Eisenhower. The majority of ship's company serving on her today was not yet born when General Eisenhower took that trip.

As part of the decom ceremony, the ship's plaque was to be hung on the sail. The ship's plaque was a ten foot long, one foot wide walnut plank with brass letters spelling out USS Seawolf. The ship's plaque could not be located on the submarine. After an anonymous suggestion, the Captain visited the Horse & Cow.

The skipper asked Jimmy, the bar owner, if he could borrow back the ship's plaque for the ceremony. He acknowledged that the plaque now belonged to the bar, having been fairly stolen and mounted on the bar's wall. Could he borrow it for the ceremony?

Jimmy said yes, provided he was invited. Invited he was, and his son Mike as well. The deal was struck and hands were shaken.

The ceremony was held pier-side on a bright spring day. Dress uniforms were everywhere, a band played. I didn't know Mare Island had a Navy band. Dignitaries we had never heard of were present. Many of them spoke at length while the crew waited at parade rest. It felt strange to have the entire crew present in one place; no one was on watch below decks.

The decommissioning ceremony over, ship's company and visitors moved to the Top Four Club. A luncheon was laid out, followed by more comments.

At a momentary lull in the speech making, Jimmy of the Horse & Cow rose from his seat. He said "I've got champagne at the Cow for *Seawolf* crew and their friends." Grateful for the escape, the club emptied. Jimmy, true to his word, had cases of chilled champagne ready. *Seawolf* sailors had kept his cash register ringing for years and he knew it. Also, he was just that kind of guy. Drinks were on the house.

The festivities over, the 'Wolf was now prepared to be towed to Bremerton, Washington. A few select crew members would travel on the towing tug to ensure her safe journey.

The crew dispersed to every corner of the Navy. Before leaving, the youngest guys passed around address books so they could keep in touch. Men who had moved more knew that this was a futile gesture. It was nice to think about but wouldn't happen. Once separated, our lives would continue to expand in different directions. People would get busy with other things. New people would move into our lives to replace the ones left behind.

Life scrolls like a computer screen. In advancing the view, new things become visible and earlier views disappear into memory. Ultimately, that too is wiped clean to make room for new.

Some people from the decommissioning crew went to separation, leaving the Navy. Some went to other submarines, some to shore duty. A few even ended up on surface craft. Perhaps eating fresh fruits and vegetables, along with seeing the sky and receiving mail, and television, and real news at sea helped them overcome the shame of skimmerhood.

Like many other former members of the 575's crew, I walked across the pier and reported for duty to USS Parche, SSN-683. Also a Projects Boat, *Parche* was newer and slated to continue *Seawolf's* work.

Before *Parche* completed her thirty years of service, she became the most decorated ship in the history of the U.S. Navy.

Once in Projects, always in Projects.

42

Monumental Terms

Submarine veterans of World War II never consider their fellow submariners 'lost.' Rather, because they went down with their ship in the service of their country and are now entombed in their final resting place beneath the sea, they and their boats are forever on 'Eternal Patrol.'

— *United States Submarine Veterans Inc.*

IN A QUIET park on a shaded corner of Submarine Base Pearl Harbor, stands a memorial dedicated to World War II submariners lost at sea. During the Second World War the submarine service sustained the highest mortality rate of all branches of the United States armed forces. One out of five U.S. Navy submariners in WWII was killed in action.

Sub Base Pearl is within a security zone and is therefore not available to the general public. Consequently the memorial is somewhat private, submariner to submariner. Fifty-two bronze plaques mounted on a simple granite wall curving gently inward. One plaque for each boat; two columns on each plaque. Three thousand, six hundred seventeen names; each one a United States Navy submariner who put to sea in the service of his country and never came home.

Contemplating death in this peaceful spot on a blue-bird afternoon seemed ridiculous. The possibility of my own death, violent or otherwise, was not real to me as a twenty-four-year-old. And yet, in all wars since God created war, it is young men who do the lion's share of the dying. The average age of those killed in the Vietnam War was twenty-three.

Now, here I was in military uniform, assigned to a nuclear submarine, a warship, soon traveling to an undisclosed location to do the undisclosed work of a covert military organization. It could be me taking first position on the list of the dead in a fresh new war as easily as someone else.

Soviet and American submarines regularly went to battle stations when the other came too close. Two big dogs, straining against their chains, snarling at each other. So far everyone had backed down before things got out of hand. So far.

Every war starts with someone being first to die. The first casualty comes as a surprise to a stunned, then outraged, nation. It comes as a surprise to KIA Number One as well. He likely was ho-humming along, going about his military business on a sunny day such as this when he came face to face with a situation that was not good. Not good at all.

The hot flash of realization. Oh shit… Then, the White Light. Now he holds first place on a list of the dead in a war he never knew.

Death is stylized in television and movies. It is of little matter that the character died; more important is how. Did he die boots-on like a man, or did he snivel? The actor dies on the screen today, but lives on a different screen tomorrow. John Wayne died in eight movies, two of them set in war. You can watch him today in those films, alive, then dead, then alive again, over and over.

Death was not real to me at twenty four. Young people consider themselves to be immortal, and everything else as non-permanent.

Teenagers make no connection between seemingly harmless actions and lasting consequences. Like the consequences of putting on a military uniform.

In every war movie a character makes a speech in which he states some version of "every man here is willing to die for the cause." The Ultimate Sacrifice. This is something live people say about those no longer alive. It's a marketing slogan to promote the military action of the day as a just cause. It's also a load of bullshit.

How it really works is frightfully simple: military men are sacrificed arbitrarily by the machinations of war. Once the machine is turned on and set in motion individuals are swept along in a tsunami of violence. If they find themselves in the wrong place at the wrong time their name goes on the dead list.

This is how I might die, so long as I wear this uniform. My life will be wagered by my government as a chess piece is moved on the board. Hopefully the cause of that wager will be just. I can deal with that. After all, I signed up and took the Man's money. I agreed to fight his wars if need be. Many of us signed on for the job and the paycheck. Many signed on for bigger reasons.

On December 8, 1941 lines of volunteers went around the block at recruiting stations throughout the country. The cause was *Rage Militare* and it was epidemic. Someone must pay for this treachery and I will see it done. A just cause and the molten steel of patriotism fired white-hot.

After the initial fury had cooled, the sustaining motivation to go was still patriotism although more measured. Patriotism reined in its pace and became duty; the long march toward victory one weary foot in front of the other. Patriotic steel was forged into weapons and will. It must be finished, and I will see it through.

In the mind-bending chaos of combat and mortal danger, military historians say what gets a man through is his sense of immediate duty, stark and clear in the moment. I must do this, right now. The greater cause doesn't exist in that moment; the situation is far too personal. Survival is all there is. The task right in front of him matters. Next month, next year, that task may seem inconsequential. In the "right here, right now" it is the only thing.

This talk of duty and sacrifice may seem like word play, but the distinctions matter. I'll bet the distinction mattered a great deal to the men whose names are on this memorial. Even though they were in a World War, with newspapers daily listing the newly dead, I wondered if the end of everything they knew seemed as unreal and far away to them as it seemed now to me. After all, they were young men too, and young men feel immortal right up until they aren't.

It wouldn't be so bad, would it, to end up listed on a stately bronze plate on a shaded wall. Then I thought of the methods by which these men's lives had ended.

Were they blown out of existence in a cataclysmic flash? Was life squeezed out of them by tons of sea pressure or the crushing air-hammer of an explosion in the confined space? Did they slowly suffocate, unable to surface their damaged ship, as the oxygen ran out and the toxic fumes of battery acid and burnt electrical gear seared their throats? Did they drown in the dark, their lungs screaming for air, overpowering the certainty that the breath they craved would kill them?

How much does it hurt to drown? Is the anguish of knowing worse than the physical pain? These grim details don't seem important when you are reading about someone else in the newspaper. "The Jones boy is dead. What a shame. He was a nice boy." The details are monumental when it could be you.

I imagined that knowing I would never again see my little daughters would cause me to die of a broken heart before the sea could kill me. When each baby girl was born I thought my heart would burst out of my chest; I had never experienced anything so sweet or so overwhelming. Imagining their tears at learning their Papa would never come home filled my eyes on the spot.

Maybe I'm not man enough to die in the service of my country.

Old infantrymen say that it is entirely random, straight chance, who walks away from war and who does not. Capability and smarts are not enough to shift the tide. They know this because they lived to become old infantrymen. Many of the companions of their youth will never be old. Many of them were smart and capable. Their names are already on monuments.

Random. Chance. Big words when applied as the cause of your end. One man lives while the next man dies. In submarines we will live or die together. Submarines do not come home with the sad tale, "we lost some guys." Everybody comes back or nobody does.

Death for a modern submariner will come in a different form than for the infantryman killed in action. It will not be bullet, hand grenade, or improvised explosive device that kills us. The mechanism by which we die will be far less individual. Most likely would be a torpedo or missile strike. It will be aimed at our boat, not at me personally. With luck, it will land on my head and in a flash I will be in the next world, pissed off at being the last to know what the hell is going on. Worst case will leave me trapped in an apocalypse of steam and radioactive boiling water, drowning in the dark, while the carcass of the boat gains speed toward an express-train crash on the sea floor 15,000 feet below.

What words would they write on our memorial plaque? Would we even get a plaque? The Cold War was a colloquialism, not a real war formally declared. The intelligence gathering done by the U.S. submarine force would fall under the category of "prudence" not "war." That would make for an impressive bronze plaque:

DIED FOR THEIR COUNTRY; JUST IN CASE.

It won't take a marvel of modern weaponry to kill a submarine. Traveling across the globe under water is a hazardous undertaking. Mother Nature is not on our side. If you doubt that, ride out a typhoon at sea. Mother Nature can kick our ass.

A submarine crew has a precarious relationship with nature. We travel immersed in nature. And yet, we can go about our business and come back alive only if we keep nature out of the People Tank.

Seawater under pressure is constantly seeking any opening to get in and cause trouble. The size of the problem will be directly proportional to the volume of water coming in and the rate at which it is doing so. An incoming trickle of seawater is an issue to deal with, to be sure, but lacking crisis. Seawater flooding in by the ton is the submarine equivalent of a massive heart attack.

Seawater flooding in can be recovered from only if we get the flooding stopped quickly enough, only if we can put on enough speed, and only if we have maintained or can create enough positive buoyancy to get us to the surface before we reach the point of no return.

Only. If. These are monumental terms.

Every watchstander must know his business and make correct choices. Every second counts. Enough speed, enough buoyancy, quickly enough. There are degrees of Enough and finally there is Not-Enough. Not-Enough lasts forever.

In April 1963, the nuclear submarine USS Thresher crossed into Not-Enough. Today she rests in 5500 feet of water, 200 miles east of Boston, Massachusetts. Every man who sailed with her that day is there still.

In May 1968, USS Scorpion found her Not-Enough. She lies silent in 10,000 feet of water southwest of the Azores along with her crew. When those capable sailors put to sea they thought they had Enough to see the day through.

A dark sense of humor is common in dangerous trades. Submariners are insatiably irreverent and joke about almost everything. However, submariners do not joke about *Thresher* or *Scorpion*. Today's submariners have a special gratitude for the men of *Thresher*. *Thresher* sank because an emergency air system failed. Directly because of that failure, a new submarine safety program began, aptly named SubSafe.

Vital systems of all U.S. submarines were redesigned, overhauled, and tested to higher standards than before to make sure that a problem like the one that killed *Thresher* never happens again. Bravo Zulu, men of *Thresher*. You paid a high price to save the rest of us.

"We" is a significant word in submarines. Long before teamwork entered the lexicon of the Human Resources department, a submarine crew was the quintessential model of it. Live together or die together. That's motivation not likely produced in a trust building seminar.

I had been standing in the Hawaiian afternoon with the submariners of WWII for a long time. My reverie was shattered by the brassy blast of a poor recording of a bugle. Played over the base public address system the familiar tune announced Evening Colors.

A short bugle call sounded "Attention." Vehicles pulled to the side of the road and stopped. Navy men and women stopped where they were, faced the flag, and stood at Attention.

As the National Anthem played over the public address system, every person in a military uniform of the United States of America rendered honors to the flag with a hand salute and held that salute.

Every flag on base and on every ship in the harbor was slowly lowered so that it reached the bottom of the flagstaff as the last note faded. Hand salutes were completed. Another short bugle call ordered "Carry On."

Vehicles rolled, work continued, people went about their business. The Color Guard carefully folded the flag into a triangle. Morning would see it raised high again.

Evening Colors lasted no more than two minutes. It was a small daily reminder of who we are and of those before us who rallied around the same flag as we and said, I will go; I will do what must be done.

I do not know what it means to die for my country. I was only beginning to know who I was when I stood in a United States Navy uniform with 3,617 submariners of World War II, and saluted the Star-Spangled Banner reverently lowered into the golden twilight of the Pacific.

We honor our country, we honor our freedom, and we honor each other. We honor those on Eternal Patrol; submariners every one, men among men.

A silver insignia, a pair of dolphin fish guiding a vessel down into the sea, hung above the left breast pocket of the last Navy uniform I wore. Those dolphins designated me Qualified in Submarines; an official member of the Silent Service, member in good standing of the Eternal Brotherhood of the Dolphin.

Once a pair of fish is nailed to your chest, you are a submariner forever.

Epilogue

United States Ship Seawolf, Submersible Ship Nuclear, hull number 575, was decommissioned at Mare Island Naval Shipyard on March 30, 1987. Towed to Bremerton, Washington, she was embalmed to await her date with the cutting torch. Her recycling officially completed on September 30, 1997. Recycled is a nice word meaning dismembered and cremated without ceremony.

When the 'Wolf was decommissioned, I transferred to USS Parche, SSN-683. Many former Wolfmen did the same. *Parche* was also home ported at Mare Island. Having grown to like Northern California, I was happy to stay.

Mare Island Naval Shipyard closed in 1996. You can freely drive on it today; no Marines, no credentials required. The island has become an industrial park and housing development.

In the summer of 2009 I drove through the unguarded gates and onto what used to be the submarine base. All around was rusting metal and buildings falling to pieces. The place looked like a post-apocalypse movie set.

I parked the car next to berth 19, former home of *Seawolf*. The only sound was the breeze flapping my shirt collar. The stillness was eerie as I recalled the activity there twenty years previous. A wave of unexpected emotion swept through me. All those guys, the wives, the girlfriends, how young we were...

Yet, life goes on. The old cemetery is still there on the eastern slope of the southern end of the island. Graves there date back to 1854. Francis Scott Key's daughter is buried there, which for a reason unknown to me, seems significant.

On the hill above the cemetery the old golf course has been remodeled and is doing great business. I wonder if those who play on those lush hillsides, overlooking the peaceful glistening of San Pablo Bay, know of the men who called that island home during the Cold War, and in service of their country went down to the sea in submersible ships.

Acronyms, Terms, and Insults

0700

Military time, the 24-hour clock. Spoken as "zero seven hundred hours."

4.0

The Navy evaluates individuals on a 4.0 scale, excellence being "four-point-oh."

575

Hull number of USS Seawolf (SSN 575).

"A" School:

Class A school. Provides technical training required to perform a job and prerequisite for more advanced training. An NEC (Naval Enlisted Classification) is awarded upon successful completion.

(SS)

Submarine Service, i.e.- Qualified in Submarines. The SS designation is written after rate and rating; e.g.- ET1(SS).

1MC

1 Main Circuit, a ship-wide announcing system. Spoken as "1MC."

2JV

General sound-powered phone system.

2MC

2 Main Circuit, propulsion plant announcing system. Spoken as "2MC."

7MC

7 Main Circuit, submarine control announcing system. Spoken as "7MC."

8K

Evaporator, distilling seawater into fresh water at a maximum rate of 8000 gallons per day .

AC

Alternating Current. Electrical current which periodically changes direction.

After Aux

After Auxiliary Machinery Room, located on USS Seawolf (SSN 575) in Stern Room Lower Level.

APD

Air Particulate Detector, detects airborne radioactive particles.

ASAP

As Soon As Possible.

ASNOB

Almost Shortest Nuke On the Boat, not yet the SNOB.

aye, aye

I understand and will obey. Followed by “Sir” to an officer.

Battery

Electrical storage device providing electrical energy to start up the reactor and to operate the submarine when the reactor is not available. A typical submarine battery consists of 126 cells, each weighing 1800 pounds and delivering 2 volts DC.

BEE

Basic Electricity and Electronics School.

Boot Camp

Euphemism for Basic Training.

Bridge

Location on the sail where the boat can be controlled on the surface. In all movies you've ever seen of the captain outside with binoculars around his neck, he is on the bridge.

BTU

British Thermal Unit, a measure of energy. The amount of energy required to raise one pound of water one degree Fahrenheit at a pressure of one atmosphere.

bulkhead

What to a landlubber would be a wall.

Captain's Mast

Non-Judicial Punishment. A formal, yet contained within the Command, short, to-the-point "trial," where the Captain is judge and jury. For lesser offenses than would merit a courts martial. A time honored maritime tradition; the captain of the ship is the highest authority.

casualty

An unplanned event, a machinery malfunction caused by enemy action or otherwise; a problem to be dealt with quickly to insure safe operation of the ship.

CDR

Navy rank of Commander, pay grade O-5.

Chief

Chief Petty Officer, pay grade E-7.

CO

Commanding Officer, referred to as "Captain" regardless of rank.

COB

Chief of the Boat, the Command Master Chief aboard a submarine, the senior enlisted man; pronounced "Cobb."

Cold War

An undeclared war for information USA vs. USSR, 1947-1991.

compartment

A submarine is comprised of watertight compartments, each capable of sealing itself with watertight doors or watertight hatches. All penetrations into and out of (electrical cables, pipes, ventilation, etc) are also watertight. Maintaining watertight integrity is vital to submarines.

COMSUBPAC

Commander Submarines Pacific

Coner

Submarine personnel whose purview is forward of the engineering spaces. On USS Seawolf (SSN 575), anything forward of Frame 57 was "forward," and anyone whose duties where forward was a Coner, or Forward Puke. Doesn't mean they're bad, they're just different than Nukes. Coners may even have an equally derisive nickname for Nukes.

Conn

Control of a ships movements; the officer in control has the Conn.

contamination

Radioactive substances where they are unintended or undesirable.

Control

Control Room, where the boat is operated from and submarine combat is engaged in.

Crew's Mess

Where the enlisted men take their meals. Other than meal time, the Crew's Mess is a place for relaxing, meeting, training, and passing time.

CTG

Coolant Turbine Generator. Steam driven turbine dedicated to powering Main Coolant Pumps.

DC

Direct Current. Electrical current traveling in one direction.

deck

Any horizontal surface you can walk on. What landlubbers would call a "floor."

decommission

Take out of service. In case of a submarine, the process takes more than a year. Every piping system is drained, every tank emptied, every useful piece of equipment that can be reused is removed. All environmental hazards are removed. The soul of the boat is sucked out and discarded. Left is only memories and maybe photos.

Det

also the Detachment; Submarine Development Group One, Detachment Mare Island.

Detailer

An anonymous person at the end of the telephone line who decides where you will go after this assignment.

dink

Delinquent in qualification progress as measured by a weekly point count. A sad state of being. Privileges will be curtailed until no longer dink.

diesel boat

A submarine powered by diesel engines which turn the screws and charge the batteries. Mostly WWII era, but some newer. Diesel Boat men have a motto DBF--Diesel Boats Forever.

diving plane

A hydraulically operated ship control surface directing the boat up and down; bow planes, stern planes. Many submarines have fairwater planes located on each side of the sail, in lieu of bow planes.

Doc

The Hospital Corpsman, the Medical Department, the guy you see when you're not feeling well or something itches.

dolphins

Informal for Submarine Warfare insignia, allowed to wear dolphins once Qualified in Submarines.

door

An opening in a bulkhead allowing passage.

DSRV

Deep Submergence Rescue Vehicle

duty section

Approximately one-quarter of the crew staying overnight to monitor and watch over the ship. Enough representatives from each division qualified in each watchstation to take the ship to sea at a moment's notice.

EAB

Emergency Air Breathing. A mask with a hose that can be plugged into manifolds located throughout the ship. One moves around from one manifold to another.

EAOS

End of Active Obligated Service. If you are exiting Naval Service, this is your independence day. If you are staying in, this is when you re-enlist.

E-DIV

Electrical Division

Emergency Diesel

Diesel Engines that power DC electrical generators to power the ship in the event of an emergency. The Diesels need air to run, therefore, the boat must be near enough to the surface to put up a mast to draw in air and vent exhaust.

ENG

Engineer Officer, also "the Engineer." Head of the Engineering Department.

Engine Order Telegraph

Electrical device on which an order is "rung up" in Control for a bell (shaft speed) that is relayed to Maneuvering. Typical shaft speeds are designated Ahead one-third, Ahead two-thirds, Ahead Standard, Ahead Full, Ahead Flank, Back one-third, Back Full, Back Emergency.

Engineering Department

Nuclear trained personnel in divisions known as Electrical (E-DIV), Mechanical (M-DIV), Reactor Controls (RC-DIV), Radiological (RL-DIV) and non-nuclear personnel in Auxiliary (A-DIV or A-gang).

Enlisted Club

On- base drinking establishment for enlisted ranks especially E-1 through E-5.

EOOW

Engineering Officer of the Watch

EPCP

Electric Plant Control Panel

EPM

Emergency Propulsion Motor

ERLL

Engine Room Lower Level

ERS

Engine Room Supervisor

ERUL

Engine Room Upper Level

Eternal Patrol

Submariners lost at sea are on "eternal patrol."

ETMS

Electronics Technician Maintenance School.

Evap-a-tron

Colloquialism for the Evaporator, a machine making potable water by distilling seawater using steam as a heat source.

EWS

Engineering Watch Supervisor

family-gram

Fifty word message a family can send to their man at sea.

Feed Station

Located in Engine Room Lower Level forward, controls secondary water into the steam generators.

Field Day

A noun and a verb; to clean the ship is to "field day" the ship. "Now commence Field Day."

fission

The reaction which takes place when a neutron is absorbed in the nucleus of a U-235 atom, upsetting the balance of energy causing the nucleus to divide, releasing energy, fission fragments, and many other particles including neutrons to produce and maintain a chain reaction.

Forward Puke

See Coner.

FUBAR

Fucked Up Beyond All Recognition

Goat Locker

Quarters for Chief Petty Officers of all grades.

GPA

Grade Point Average.

Great Lakes

Great Lakes Naval Training Center, GLNTC, located north of Chicago, Illinois.

Green Band

Optimum Average Primary Coolant temperature range controlled the Reactor Operator. The meter on the Reactor Plant Control Panel has this range marked in green.

H&C

The Horse & Cow Bar, also The Cow; a safe haven for submariners in Vallejo, California. Now gone, the original owners have started other locations.

Halfway Night

A ritual evening of festivities in the Crew's Mess at the halfway point in a mission.

hatch

A closable device for passage up and down.

head

Navy-talk for toilet.

Hollywood shower

A bathing ritual consisting of standing under hot water for a prolonged term which is not approved for shipboard use. A Navy shower properly done water on, get wet, water off, soap up, water on, rinse, water off.

JO

Junior Officer; Ensign, Lieutenant Junior Grade, Lieutenant.

ladder

Allows up and down passage, sometimes looks like a ladder, often resembles stairs. Sailors use ladders, landlubbers use stairs.

LCDR

Lieutenant Commander; pay grade O-5

LCPO

Leading Chief Petty Officer

liberty

Authorized absence. Time off the boat.

LMET

Leadership and Management Educational Training

LPO

Leading Petty Officer

LT

Lieutenant; pay grade O-3

LTJG

Lieutenant junior grade; pay grade O-2

MAA

Master-at-Arms, Ships' police. Large ships and shore facilities have a dedicated military police force, submarines do not.

Main Engines

Main Engines, driven by propulsion turbines.

Maneuvering

Maneuvering Room, control station for the reactor plant, electric plant, and propulsion. Watch station for Engineering Officer of the Watch (EOOW), Reactor Operator (RO), Electrical Operator (EO), and Throttleman.

Maneuvering Watch

Entering or leaving port where precise ship control is critical, stationing the Maneuvering Watch puts the most senior or skilled watchstanders on duty along with extra watchstanders with specific responsibilities.

Mare Island

Mare Island Naval Shipyard, located near Vallejo, California.

M-DIV

Machinery Division

MINSY

Acronym for Mare Island Naval Shipyard.

MOO

Master of the Obvious

MOOOC

Master of the Obvious Out Of Control

MPA

Main Propulsion Assistant. Division officer for Machinery Division.

Nautilus
USS Nautilus (SSN 571), the first nuclear-powered submarine, launched September 30, 1954.

Navigator
Officer who is head of the navigation department. Responsible for navigation of the ship.

NEC
Naval Enlisted Classification. A code describing a skill set.

NFQT
Nuclear Field Qualification Test.

NIs
Nuclear Instruments. Monitor reactor power level.

NJP
Non Judicial Punishment. Handed out at Captain's Mast.

NPS
Nuclear Power School.

NPTU
see Nuclear Power Training Unit

NRRO
Naval Reactors Regional Office. The local Naval Reactor representatives, the bishops of Hyman G. Rickover.

Nuclear Power Training Unit
Nuclear Power Training Unit, NPTU, the back end of a submarine on land, used to train Navy Nukes.

Nucleonics
Laboratory for Engineering Laboratory Technicians.

Nuke
Nuclear Trained personnel.

Officer's Club
Drinking and dining venue for officers only.

OOD
Officer Of the Deck. Captain's representative controlling the ship, on watch in Control.

ORSE

Operational Reactor Safeguards Examination.

overhead

What to a landlubber would be a “ceiling.” A broader term, most anything above you is overhead or "in the overhead."

P&A

Protection & Alarm system, designed to protect the reactor from damage. Monitors primary plant parameters and automatically shuts down the reactor if necessary.

Parche

USS Parche (SSN-683)

Petty Officer

Non-commissioned officer in grades of master chief, senior chief, chief, and first, second, and third class.

Plank Owner

Person who was onboard a ship at commissioning or decommissioning.

PFM

Pure Fucking Magic. The simplest way to describ the operation of complex electronic circuitry to those unfamiliar to complex electronic circuitry.

PMS

Preventive Maintenance Schedule, a comprehensive scheduling system for preventive maintenance on every piece of equipment on a ship.

POD

Plan of the Day

port

Left side when facing forward, right side is starboard. Also a seaside city open to ships.

PPIP

Primary Plant Instruments Panel.

Prac Fac

Practical Factor, part of watchstation or ship's qualification where the candidate accomplishes something physical under the supervision of a qualified person.

Primary Sample Sink

Provides limited access to primary plant water for chemistry samples.

Projects

Projects Department; mystical magical entity and compartment on Special Projects submarine where Project Weenies do what they do.

Prototype

see Nuclear Power Training Unit

Qual Standard

Qualification Standard. Details specific level of knowledge to be demonstrated before being formally qualified in a watch station or certified in a field as in the nuclear field.

qualify

The process to become formally recognized as having the knowledge and skills to perform a specific watchstation or task.

qualify Ship's

To qualify on the submarine. Demonstrate knowledge of all ship's systems, both theory and operation of.

rack

Submariner-speak for bunk or bed.

radiation

General term applied to emission of radioactive particles. These particles come in many forms, some more dangerous than others.

RC

Reactor Compartment

RC-DIV

Reactor Controls Division

RCPO

Recruit Chief Petty Officer. A faux title and authority used in Navy Basic Training.

Reactor Compartment Lower Level

On USS Seawolf the space containing the reactor plant, inaccessible during operation.

Reactor Compartment Upper Level

On USS Seawolf the space containing reactor control equipment, the work space for Reactor Controls Division.

Reduction Gears

Gear system that converts the high speed of the propulsion turbines to the slower speed of the propeller shaft.

RL-DIV

Radiological Division

RPCP

Reactor Plant Control Panel

RPM

Reactor Plant Manual. A multi-volume set of manuals providing procedures for operation, maintenance, and casualty control of the reactor plant and propulsion plant.

rpm

Revolutions per minute

RT

Reactor Technician. A watchstation outside of Maneuvering to monitor and operate reactor controls equipment.

rudder

Ship control surface located at the stern of a vessel providing right and left directional control. Operated in Control by the Helmsman.

Russell

USS Richard B. Russell (SSN 697), a SUBDEVGRUONE boat home ported at Mare Island.

S1C

Nuclear Power Training Unit, located in Windsor Locks, Connecticut. Also designation of reactor plant; Submarine reactor, model 1, built by Combustion Engineering.

S2Wa

Designation of USS Seawolf (SSN 575) reactor plant; Submarine reactor, model 2, built by Westinghouse, modified.

S5W

Designation of widely used reactor plant; Submarine reactor, model 5, built by Westinghouse.

SAN2

Sanitary Tank number 2, storage tank for toilet, washroom, and shower drains.

SAT

Satisfactory, a desirable state of being for anything Navy. Meeting or exceeding the standard.

Scorpion

USS Scorpion (SSN 589), lost at sea 22 May, 1968 with 99 men aboard.

SCRAM

A rapid reactor shutdown procedure accomplished automatically by the reactor Protection & Alarm system or manually by the Reactor Operator or Reactor Technician.

screw

Colloquialism for the ship's propeller.

sea-dad

A mentor assigned to a new report.

Seawolf

USS Seawolf (SSN 575), the second nuclear-powered submarine, commissioned 30 May 1957, decommissioned 30 May 1987. A SUBDEVGRUONE boat home ported at Mare Island.

Secondary Sample Sink

Provides access to secondary plant (steam plant) water for chemistry samples.

Ship's Office

Pay, leave, orders, awards, the POD, and all manner of official documents and decrees emanate from the Ship's Office.

Shore Patrol

Naval police force for shore installations.

skimmer

A surface craft or a person assigned to surface craft. See also Target.

SNAFU

System Normal; All Fucked Up.

SNOB

Shortest Nuke On the Boat. The nuclear trained person with the least time remaining to End of Active Obligated Service (EAOS).

sound-powered telephone

Ship-wide interior communication system powered only by the voice, no outside source of electricity required. When everything else goes dark, the sound-powered phones will work. The 2JV circuit.

soup sandwich

The ultimate adsurdity.

SPCP

Steam Plant Control Panel, a.k.a.- the Throttles.

Special Projects

Refers to the Office of Special Projects or the Special Projects department on the submarine.

SRO

Shutdown Reactor Operator. Watchstation in Maneuvering manned 24/7 while the reactor is shutdown.

SSTG

Ship's Service Turbine Generator. Steam driven turbine turns an electrical generator mounted on the same shaft. Makes electricity and lots of it.

staff pickup

A NPTU graduate asked to stay on as a staff instructor.

starboard

Right side when facing forward, left side is port.

STE

Shift Test Engineer. Civilian who works for thc shipyard, as liaison between sailors and civilians for shipyard work and testing while "in the Yards."

SUBDEVGRUONE

Submarine Development Group One

submariner

A man among men, a Steely-Eyed Killer of the Deep.

target

Any surface vessel. See also skimmer.

Thresher

USS Thresher (SSN 593); lost at sea 10 April 1963 with 129 men.

TLD

Thermoluminescent dosimeter. A personal radiation monitoring device worn on the belt.

Top Four Club

A shore drinking and dining establishment reserved for the top four levels of enlisted; E6-E9.

TS-SBI

Top Secret with a Special Background Investigation, the security level required for Special Projects boats.

Torpedo Room

Forward most compartment of USS Seawolf (SSN 575), where torpedoes are stored, maintained, and loaded into torpedo tubes for firing. Also used for crew berthing.

training

A daily ritual on submarines at many levels Ship's training, officer's training, departmental training, divisional training, individual quals training, individual training in rate.

Triton

USS Triton (SSN 586), first submarine to circumnavigate the earth submerged, 1960

UA

Unauthorized Absence.

UCMJ

Uniform Code of Military Justice.

UI

Under Instruction. A UI watch is a student watchstander under the direction of a qualified watchstander and always within reach. The qualified watchstander trains the UI and prevents the UI watch from making mistakes.

UNSAT

Unsatisfactory. An undesirable state of being requiring remediation, adjustment, or repair. Applies to machinery and personnel.

USN

United States Navy.

Usta-boat

Any previous submarine on which one has been stationed.

Wardroom

Meeting, dining, studying room for officers only. Enlisted allowed only on official business.

watch

A Navy term with many iterations. A specific assigned duty or the person manning that station. The Reactor Operator is a watch station (place), assigned to a specific qualified reactor operator (person) for six hours (time period), on a regular rotation at sea, e.g.-the Midwatch 0000-0600 (a specific reccurring time period). When looking for the Topside Watch, you are looking for the person standing duty as the Topside Watch. If you have the Topside Watch, they are looking for you. When you relieve the watch, you now have the watch and he does not have the watch any longer.

working party

An *ad hoc* group performing a task, e.g.- a stores load working party; transferring food from trucks on the dock to the interior of the submarine.

XO

The Executive Officer, second in command; referred to as "the Executive Officer," or "the XO," or "XO."

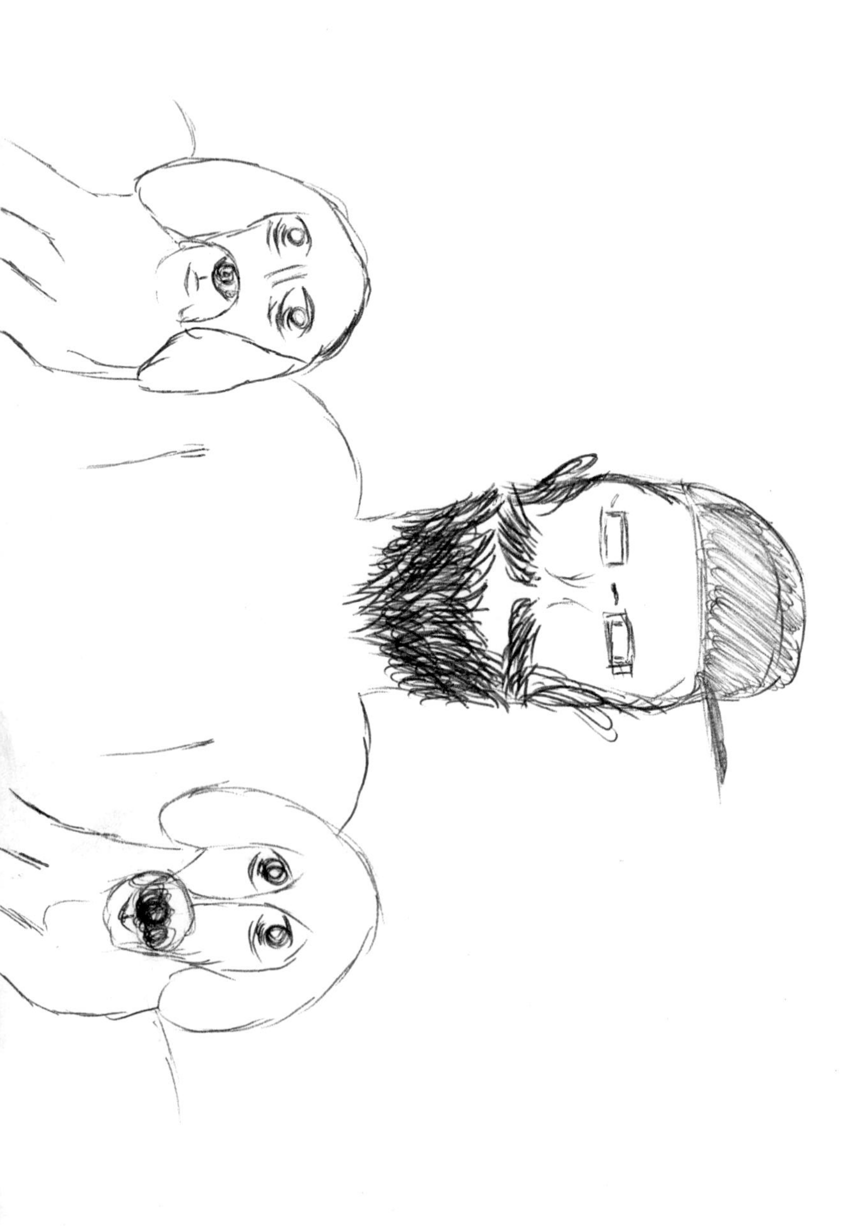